The Sand Compaction Pile Method

The Sand Compaction Pile Method

Masaki Kitazume, Dr. Eng.

Port and Airport Research Institute,
Yokosuka, Japan

CRC Press
Taylor & Francis Group
Boca Raton   London   New York

CRC Press is an imprint of the
Taylor & Francis Group, an **informa** business

A BALKEMA BOOK

Contents

CHAPTER 3   DESIGN PROCEDURES FOR SANDY GROUND

Preface

Due to the growing population and spreading urbanization in the past century, it has become difficult to obtain suitable ground for constructing infrastructures. Construction projects often encounter very soft soil deposits, which can pose problems of stability, excessive settlement and/or liquefaction. To solve these problems, a variety of ground improvement techniques have been developed and put into practice so far. The Sand Compaction Pile Method (SCP), in which many compacted sand piles are constructed, was developed for improving clay and sandy grounds in Japan. Numerous research efforts have been made in Japan to investigate the shear and consolidation characteristics of composite soil with sand piles and clay, deformation and failure behavior of SCP improved ground under static and dynamic conditions, design method, and execution techniques. At present, each year several thousand kilometers of SCP improvement are done in Japan. Similar techniques such as the stone column method and vibro-flotation method have been developed in foreign countries. As there have been limited occasions to exchange information, these methods have been developed individually. Recently, the SCP method has been adopted in many foreign countries especially in Asia for improving sandy and clay grounds.

More than 500 research papers on many aspects of the SCP method have been published in international and domestic conferences and journals. Unfortunately, as many research studies are written in Japanese, accumulated research efforts, practical experience and know-how have not been widely disseminated in foreign countries. This book aims to introduce research and development, various applications, and the design and execution of the SCP method to other countries, based on the extensive research efforts and accumulated experience in Japan. As soils are local materials and the applications are different in various parts of the world, the book is not intended to provide a complete design manual for the SCP method; rather, it shows the state of practice of the technology. In actual design and construction, it is always advisable to employ the latest engineering expertise and know-how.

Many figures and tables cited in this text are originally written in Japanese. In preparing the text, these figures and tables are translated into English by me. Where possible, the different unit systems adopted in the original research papers have been converted into the SI unit system and specific figures and numbering are also unified for ease of reading.

I hope that this book will be a useful reference for practical engineers, research engineers and academics interested in the Sand Compaction Pile Method around the world.

November 2004

Masaki Kitazume

Head of Soil Stabilization Division
Geotechnical and Structural Engineering Department
Port and Airport Research Institute

List of Technical Terms and Symbols

a_s	replacement area ratio
a_{eq}	equivalent acceleration (gal)
a_{max}	maximum acceleration (gal)
A	cross sectional area of clay ground and sand pile (m^2)
A_c	cross sectional area of clay ground (m^2)
A_s	cross sectional area of sand pile (m^2)
B	width of improved ground (m)
B	width of foundation (m)
B	diameter or width of pile structure (m)
c	undrained shear strength of clay between sand piles (kN/m^2)
Cc	compressive coefficient
C_k	correction factor of input motion
c_0	cohesion of clay at ground surface (kN/m^2)
c_u	undrained shear strength of clay ground (kN/m^2)
c_u/p	shear strength increment ratio
d	diameter of sand pile (m)
D	interval of sand piles (m)
D	embedded depth of foundation (m)
D	depth of improved ground, (m)
D_r	relative density (%)
e_i	initial void ratio
e_{max}	maximum void ratio
e_{min}	minimum void ratio
EI	flexural rigidity of pile structure (kN m^2)
Fc	fines content (%)
Fs	safety factor
H	ground thickness (m)
k	increment ratio of shear strength of clay with depth (kN/m^3)
k	coefficient of seismic intensity
k_h	coefficient of subgrade reaction (m)
k_{hc}	coefficient of subgrade reaction of clay ground (kN/m^3)
k_{hs}	coefficient of subgrade reaction of sand pile (kN/m^3)
k_{h0}	coefficient of subgrade reaction of ground at 1 cm horizontal displacement of 1 cm diameter pile (kN/m^3)

k_{hc0}	coefficient of subgrade reaction of clay ground at 1 cm horizontal displacement of 1 cm diameter pile (kN/m^3)
k_{hs0}	coefficient of subgrade reaction of sandy ground at 1 cm horizontal displacement of 1 cm diameter pile (kN/m^3)
K_0	static earth pressure coefficient
m_v	volumetric coefficient (m^2/kN)
n	stress concentration ratio
N	SPT N-value
N_1	normalized SPT N-value with respective overburden pressure of 98 kN/m^2 (1 kgf/cm^2)
N_{65}	equivalent SPT N-value
N_i	SPT N-value of original ground
N_m	measured SPT N-value
N_t	SPT N-value of improved ground incorporating fines content effect
N_t	SPT N-value of improved ground without incorporating fines content effect
N_{ti}	SPT N-value at sandy ground between sand piles
N_{ts}	SPT N-value at center of sand pile
N_c	bearing capacity factor
N_q	bearing capacity factor
N_γ	bearing capacity factor
P	subgrade reaction force acting on pile structure (kN/m)
p	subgrade reaction pressure acting on pile structure (kN/m^2)
P	bearing capacity (kN)
P_A	active earth pressure of improved ground (kN/m^2)
P_P	passive earth pressure of improved ground (kN/m^2)
p_{ws}	static water pressure (kN/m^2)
p_{wd}	dynamic water pressure (kN/m^2)
q_a	bearing capacity of improved ground (kN/m^2)
q_{ac}	bearing capacity of clay ground (kN/m^2)
q_{as}	bearing capacity of sandy ground (kN/m^2)
q_u	unconfined compressive strength (kN/m^2)
Rc	effective compaction ratio
R_{max}	*in-situ* liquefaction resistance
S	settlement of improved ground (m)
S_0	settlement of original (unimproved) ground (m)
S_t	settlement of improved layer (m)
S_u	settlement of unimproved layer underlying SCP improved layer (m)
U	average degree of consolidation
U_c	uniformity coefficient
α	maximum acceleration at ground surface (gal)
β	settlement reduction factor
β	reduction factor
$1/\beta$	characteristic length (m)

γ_c unit weight of clay (kN/m^3)

γ_s unit weight of sand pile (kN/m^3)

γ_m average unit weight of improved ground (kN/m^3)

γ' effective unit weight (kN/m^3)

γ_{sat} saturated unit weight (kN/m^3)

ΔN_f increment of SPT N-value for fines content effect

ΔN increment of SPT N-value – actual

$\Delta N'$ increment of SPT N-value – predicted

μ coefficient of upheaval (%)

μ_c stress concentration coefficient of clay ground for external load

μ_s stress concentration coefficient of sand pile for external load

σ vertical stress (kN/m^2)

σ_0 initial vertical stress (kN/m^2)

σ_c vertical stress on clay ground (kN/m^2)

σ_h horizontal stress (kN/m^2)

σ_s vertical stress on sand pile (kN/m^2)

σ_u upper yield stress (kN/m^2)

σ_v vertical stress (kN/m^2)

σ_v' effective overburden pressure (kN/m^2)

τ shear strength (kN/m^2)

τ_{max} maximum shear stress (kN/m^2)

ϕ internal friction angle

ϕ_m internal friction angle of improved ground

ϕ_s internal friction angle of sand pile

Chapter 1

Outline of the Sand Compaction Pile Method

1.1 INTRODUCTION

It is an obvious truism that structures should be constructed on good quality ground. Ground conditions of construction sites, however, have become worse in recent decades throughout the world. This situation is especially pronounced in Japan, where many construction projects are conducted on soft alluvial clay grounds, reclaimed grounds with dredged soils, highly organic soil grounds, loose sandy grounds and so on. When any types of infrastructures are constructed large amounts of ground settlement and/or stability failure are likely to be encountered. Apart from clay or highly organic soil grounds, loose sand deposits under a water table cause serious problems of liquefaction under seismic conditions. In such cases, suitable soil improvement techniques are required to improve the physical properties of soft soil in order to cope with these problems. Many soil improvement techniques have been developed in Japan and other countries for these purposes.

The Sand Compaction Pile (SCP) method has been developed and frequently adopted for many construction projects in Japan, in which sand is fed into a ground through a casing pipe and is compacted by either vibration, dynamic impact or static excitation to construct a compacted sand pile in a soft soil ground. This method was originally developed in order to increase the density of loose sandy ground and to increase the uniformity of sandy ground, to improve its stability or compressibility and/or to prevent liquefaction failure, but now it has also been applied to soft clay ground to assure stability and/or to reduce ground settlement. The principal concept of the SCP method for application to sandy ground is to increase the ground density by placing a certain amount of granular material (usually sand) in the ground. The principal concept for application to clay ground, on the other hand, is reinforcement of composite ground consisting of compacted sand piles and surrounding clay, which is different from that of the Sand Drain method in which sand piles without any compaction are constructed principally for drainage function alone.

According to improvement principles, soil improvement methods can be categorized into five groups: replacement, consolidation, densification, solidification and contact pressure reduction, as shown in Figure 1.1 [1]. The figure summarizes many soil improvement techniques available in Japan. Among the categories, the SCP method can be categorized into 'reinforcement method' and 'densification method' for applications to clay and sandy grounds respectively. Figure 1.1 also shows the period in which practical application of each method was introduced. The SCP method was practically applied as a densification method in 1957 and as a replacement method in

Improvement principle	Engineering method	Work examples	Period practical application introduced							
			1930s	1940s	1950s	1960s	1970s	1980s	1990s	2000s
Replacement	Excavation replacement	Dredging replacement method	⊢————————————————————————▶							
	Forced replacement	Sand compaction pile method				1966 ⊢——————————▶				
Consolidation	Preloading	Preloading method	1928 ⊢————————————————————▶							
	Preloading with vertical drain	Sand drain method			1952 ⊢——————————————▶					
		Packed sand drain method				1967 ⊢——————————▶				
		Board drain method				1963 ⊢——————————▶				
	Dewatering	Deep well method		1944 ⊢————————————————▶						
		Well point method			1953 ⊢——————————————▶					
		Vacuum consolidation method					1971 ⊢——————▶			
	Chemical dewatering	Quick lime pile method				1963 ⊢——————————▶				
Densification — Dewatering/compaction	Compaction by displacement and vibration	Sand compaction pile method			1957 ⊢——————————————▶					
		Gravel compaction pile method				1965 ⊢——————————▶				
Densification — Compaction	Vibration compaction	Vibro-flotation method			1955 ⊢——————————————▶					
	Impact compaction	Dynamic consolidation method					1973 ⊢——————▶			
Solidification (Admixture stabilization)	Agitation mixing	Shallow mixing method					1972 ⊢——————▶			
		Deep mixing method					1974 ⊢——————▶			
	Jet mixing	Jet mixing method						1981 ⊢——▶		
Contact pressure reduction	Load distribution	Fascine mattress method	⊢————————————————————————▶							
		Sheet net method				1962 ⊢——————————▶				
		Sand mat method	⊢————————————————————————▶							
		Surface solidification method					1970 ⊢——————▶			
	Balancing loads	Counterweight fill method	⊢————————————————————————▶							

Figure 1.1.   Principle of soft ground improvement methods in Japan [1].

1966, which was rather early among the improvement methods. This category includes other sand pile constructions methods such as 'vibro-compaction' for sandy grounds and 'vibro-replacement' and 'vibro-displacement' for clay grounds, which have been applied in foreign countries. The procedure of constructing compacted sand piles in these methods is somewhat different from the SCP method: the sand piles in these methods are constructed by feeding sand in the cavity outer surface of the casing pipe. However the shape and function of compacted sand piles in a ground are similar to those of the SCP method.

This text describes the design, execution, quality control and assurance of the SCP method in detail together with some applications.

1.2   IMPROVEMENT PURPOSE AND APPLICATIONS

In Japan, the Sand Compaction Pile (SCP) method has been applied to improve soft clays, organic soils and loose sandy soils for various purposes and in various ground conditions. Figure 1.2 shows typical improvement purposes of the SCP method in Japan. The improvement purposes can be divided into two categories: clay ground and sandy ground. Applications to clay ground include increasing bearing capacity, reducing settlement, increasing passive earth pressure, reducing active earth

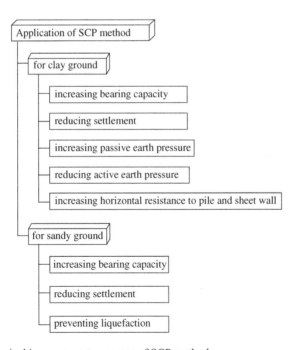

Figure 1.2.   Typical improvement purposes of SCP method.

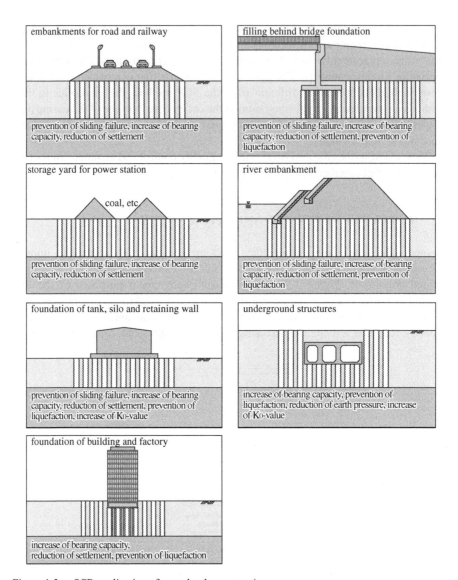

Figure 1.3.   SCP applications for on-land construction.

pressure and increasing horizontal resistance to pile and sheet wall. Applications to sandy ground, on the other hand, include: increasing bearing capacity, reducing settlement and preventing liquefaction.

Figures 1.3 and 1.4 show typical applications of the SCP method to on-land construction and marine construction respectively. In on-land construction, the SCP method has been applied to embankments, oil tanks, and building foundations, while

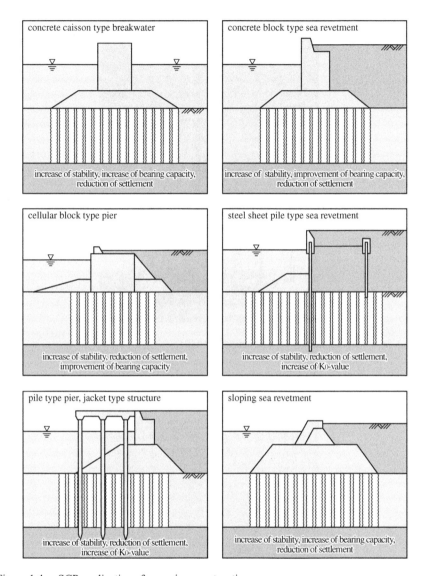

Figure 1.4.   SCP applications for marine construction.

the SCP method has been applied to breakwaters, sea revetments and piers in marine construction.

Figures 1.5a and 1.5b show typical SCP machines for on-land construction and marine construction respectively. In the method, many compacted sand piles are constructed in a ground with square, equilateral triangular or rectangular pattern by SCP machines.

Figure 1.5a.   SCP machine for on-land construction [2].

Figure 1.5b.   SCP barge for marine construction [2].

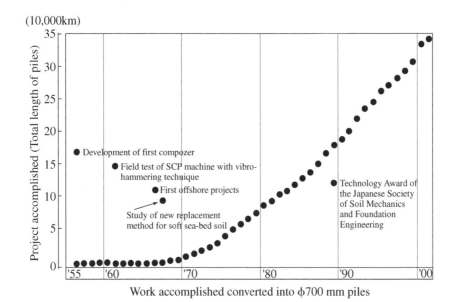

Figure 1.6.   Cumulative length of compacted sand piles [2].

According to the wide varieties of SCP application together with the development of SCP machines, the maximum improvement depth of the method increased to 70 m in 1993. The cumulative length of compacted sand piles also increased very rapidly in the 1970s, 1980s and 1990s, and reached 350 thousand km in 2001, as summarized in Figure 1.6 [2]. Nowadays, the SCP method has been used to improve many kinds of ground including clay ground, sandy ground and fly ash ground for various improvement purposes.

1.3   SCOPE OF TEXT

This text is not intended to be a complete design manual for the SCP method, but to give the latest state of practice and the state of the art on the design, construction and quality control and assurance of the SCP method.

In Chapters 1 to 4, the design, construction and QC/QA of the method are described in detail so that geotechnical students and practical engineers can easily understand and adopt the SCP method to their studies and construction projects. Among the many applications, some case histories are briefly introduced in Chapter 5, which include several kinds of original ground condition, purpose of improvement, SCP material and machine. These case histories clearly show a wide variety of SCP applications. In Chapters 6 to 9, on the other hand, the historical background of development of the SCP method and important issues to be considered for more sophisticated design are described in detail. The state of the art reports on the strength and deformation characteristics of SCP improved ground are also described in detail in Chapter 7. These chapters provide detailed, thorough information and accumulated research and experiences for academics and senior research engineers as useful references for their research.

As the sand piles in the 'stone column method', 'vibro-compaction', 'vibro-replacement' and 'vibro-displacement' methods are similar to those of the SCP method, the basic concept of the design and quality control and assurance described in this text can be applicable to these methods.

REFERENCES

1   Ministry of Transport: Technical standards and commentaries for port and harbour facilities in Japan. The Japan Ports and Harbours Association, 1999 (in Japanese).
2   The courtesy of Fudo Construction Co. Ltd.

Chapter 2

Design Procedures for Clay Ground

2.1 INTRODUCTION

In the SCP method, many compacted sand piles are constructed by a series of construction operations using SCP machines. A group column type improvement pattern has usually been applied in order to increase bearing capacity and stability, and to reduce vertical and horizontal displacement. In this chapter, several current design procedures for clay grounds are described in detail, while those for sandy grounds will be described in the next chapter. The design procedures introduced in the following sections are still being modified in accordance with new findings from many research projects and accumulated experiences, which include static and dynamic behaviors of improved ground, and a great deal of know-how in design and practice. Therefore a specific practical design for each construction site should be carried out not only according to the procedure, but also in accordance with ongoing research activities and accumulated know-how.

2.2 IMPROVEMENT PATTERNS AND STRESS CONCENTRATION RATIO

2.2.1 Improvement pattern

In Japan, the Sand Compaction Pile (SCP) method has been applied to improve soft clays, organic soils and loose sandy soils for various purposes and in various ground conditions. For these improvement purposes, many compacted sand piles are constructed in grounds with square, equilateral triangular or rectangular pattern, as shown in Figure 2.1. The replacement area ratio, a_s, is defined as the ratio of the sectional area of the sand pile to the hypothetical cylindrical area, which is formulated as Equation (2.1) for square, equilateral triangular and rectangular pile installation patterns (see Figure 2.2). The replacement area ratio for applications to sandy ground is typically less than 0.3, and for clay ground the ratio ranges from 0.3 to 0.8. In the case of the replacement area ratio of about 0.78, sand piles are in contact with each other. The SCP improved ground can be divided into 'low', 'medium' and 'high' for a replacement area ratio of less than 0.3, 0.3 to 0.5, and higher than 0.5.

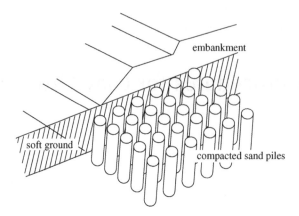

Figure 2.1.   Group column type improvement.

Replacement area ratio:

$$as = \frac{A_s}{A}$$
(2.1)

for square patterns:

$$as = \frac{A_s}{A} = \frac{\pi d^2/4}{D^2}$$
(2.1a)

for equilateral triangular patterns:

$$as = \frac{A_s}{A} = \frac{\sqrt{3}}{2} \cdot \frac{\pi d^2/4}{D^2}$$
(2.1b)

for rectangular patterns:

$$as = \frac{A_s}{A} = \frac{\pi d^2/4}{D_1 \cdot D_2 \cdot \sin \theta}$$
(2.1c)

where:
- as : replacement area ratio
- A : cross sectional area of clay ground and sand pile (m^2)
- A_s : sectional area of sand pile (m^2)
- d : diameter of sand pile (m)
- D : interval of sand piles (m)
- D_1 : interval of sand piles for rectangular pattern (m) (see Figure 2.2(c))
- D_2 : interval of sand piles for rectangular pattern (m) (see Figure 2.2(c))
- θ : angle of sand piles arrangement for rectangular pattern

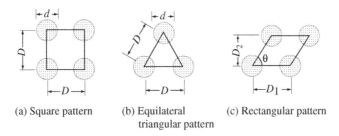

<table>
<tr><td>(a) Square pattern</td><td>(b) Equilateral
triangular pattern</td><td>(c) Rectangular pattern</td></tr>
</table>

Figure 2.2.   Improvement pattern.

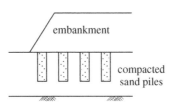

Figure 2.3a.   Fixed type improvement.

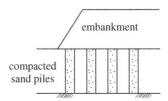

Figure 2.3b.   Floating type improvement.

2.2.2   Improvement type

SCP improved ground can also be divided into two improvement types as schematically shown in Figures 2.3a and 2.3b: fixed and floating types depending upon whether sand piles reach a stiff bearing layer or not. The fixed type improvement is a type of improvement where sand piles reach a stiff layer, while the floating type improvement is where sand piles do not reach a stiff layer but end in a soft soil layer. It can be easily understood that the fixed type improvement is preferable from the viewpoints of increasing stability and reducing settlement. In the case where the thickness of the soft layer is quite large, however, the floating type improvement is recommended.

2.2.3   Stress concentration ratio

A clay ground improved by the SCP method can be considered to be a composite ground consisting of compacted sand piles and surrounding clay as schematically shown in Figure 2.1. External load is concentrated mainly on the sand piles as shown in Figure 2.4, because the compressive stiffness of sand piles is much higher than that of the surrounding clay ground, while less load is applied to the clay between the sand piles. The stress concentration ratio, n, is defined as the ratio of the vertical stresses acting on sand piles, σ_s, to that on the surrounding clay, σ_c. A simple analytical approach provides a formulation of these stresses as expressed in Equations (2.2) to (2.6), which are derived from the stress equilibrium between sand piles and surrounding clay.

(1) The vertical stress:

$$\sigma \cdot A = \sigma \cdot (A_c + A_s)$$
$$= \sigma_c \cdot A_c + \sigma_s \cdot A_s \qquad (2.2)$$

(2) when introducing the stress concentration ratio:

$$n = \frac{\sigma_s}{\sigma_c} \qquad (2.3)$$

(3) and the replacement area ratio (see Equation (2.1)):

$$as = \frac{A_s}{A_c + A_s} \qquad (2.4)$$

(4) the vertical stress on clay ground can be expressed as:

$$\sigma_c = \frac{\sigma}{\{1 + (n - 1) \cdot as\}} \qquad (2.5)$$

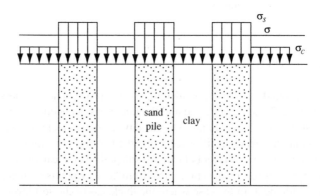

Figure 2.4.   Illustration of stress concentration.

(5) the vertical stress on sandy ground can be expressed as:

$$\sigma_s = \frac{n \cdot \sigma}{\{1 + (n - 1) \cdot as\}} \tag{2.6}$$

where:

  as : replacement area ratio
  A : cross sectional area of clay ground and sand pile (m^2)
  A_c : cross sectional area of clay ground (m^2)
  A_s : cross sectional area of sand pile (m^2)
  n : stress concentration ratio
  σ : average vertical stress (kN/m^2)
  σ_c : vertical stress on clay ground (kN/m^2)
  σ_s : vertical stress on sand pile (kN/m^2)

2.3   BEARING CAPACITY

2.3.1   Bearing capacity of isolated sand piles

In the case where sand piles are installed with relatively large interval and/or the sectional area of loading is quite small with respect to the sand pile diameter, the bearing capacity of an isolated sand pile should be evaluated. The bearing capacity of a single sand pile and surrounding soil, P, was proposed as Equations (2.7) to (2.9) by Murayama [1], in which the stress equilibrium of sand piles and surrounding soil, and the stress concentration effect are incorporated (see Figure 2.5). The principle of his calculation is that the sand piles and surrounding clay are subjected to equal vertical settlement, leading to a stress concentration on the sand piles. In this formulation, the

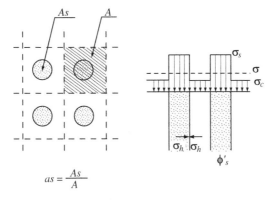

Figure 2.5.   Illustration of bearing capacity of a single sand pile.

stress conditions of sand pile and surrounding soil are assumed as an ultimate active state and an ultimate passive state respectively.

(1) The vertical stress:

$$
\begin{aligned}
P &= A \cdot \sigma \\
&= (A_s + A_c) \cdot \sigma \\
&= A_s \cdot \sigma_s + A_c \cdot \sigma_c \\
&= \sigma_c (A_s \cdot n + A_c)
\end{aligned}
$$
(2.7)

(2) the horizontal stress should be satisfied the following equation:

$$
\sigma_h \geq \frac{1 - \sin \phi_s}{1 + \sin \phi_s} \cdot \sigma_s
$$
(2.8a)

$$
\sigma_h \leq \sigma_c + \sigma_u
$$
(2.8b)

(3) the stress concentration ratio can be expressed as:

$$
\frac{\sigma_s}{\sigma_c} = n
$$

$$
= \frac{1 + \sin \phi_s}{1 - \sin \phi_s} \cdot \left(1 + \frac{\sigma_u}{\sigma_c}\right)
$$
(2.9)

where:
  a_s : replacement area ratio
  A : cross sectional area of clay ground and sand pile (m²)
  A_c : cross sectional area of clay ground (m²)
  A_s : cross sectional area of sand pile (m²)
  n : stress concentration ratio
  P : bearing capacity (kN)
  σ : average vertical stress (kN/m²)
  σ_c : vertical stress on clay ground (kN/m²)
  σ_s : vertical stress on sand pile (kN/m²)
  σ_h : horizontal stress on cylindrical surface of sand pile (kN/m²)
  σ_u : upper yield stress of clay ground (kN/m²)
  ϕ_s : internal friction angle of sand pile

After substituting Equations (2.8) and (2.9) into Equation (2.7), the bearing capacity, P, is calculated by Equation (2.10):

$$
P = \sigma_u \cdot \frac{1 + \sin \phi_s}{(n - 1) + (n + 1) \cdot \sin \phi_s} \cdot (A_s \cdot n + A_c)
$$
(2.10)

The upper yield stress of clay ground, σ_u, can be estimated as $2c_u = q_u$, where c_u and q_u are undrained shear strength and unconfined compressive strength of surrounding clay respectively, if the clay can be assumed to fail by an ultimate passive condition.

However, Murayama [1] assumed that σ_u is $0.7q_u$ by considering that no infinite creep deformation took place in the clay. By substituting this value into Equation (2.10), finally the bearing capacity, P, can be obtained as Equation (2.11):

$$P = 0.7 \cdot q_u \cdot \frac{1 + \sin \phi_s}{(n - 1) + (n + 1) \cdot \sin \phi_s} \cdot (A_s \cdot n + A_c) \qquad (2.11)$$

where:
   q_u : unconfined compressive strength of clay ground (kN/m^2)

2.3.2   Bearing capacity of improved ground with multi sand piles

In the case where the interval of sand piles is not large enough to be isolated, and/or the structure size is large enough to be supported by many sand piles, an interaction of sand piles and surrounding clay should be incorporated in evaluating the bearing capacity. The bearing capacity in this case can also be calculated by Equation (2.11) provided the upper yield stress of clay ground, σ_u, can be precisely estimated. However, this has not yet been clarified under an interactive situation of many sand piles. For this case, the bearing capacity of SCP improved ground is often evaluated by Terzaghi's bearing capacity theory as shown in Equation (2.12) [2] or slip circle analysis. The bearing capacity evaluation by Terzaghi's theory is described here, while that by the slip circle analysis will be described in the next section.

   In the bearing capacity evaluation by Terzaghi's theory, the bearing capacities of uniform clay ground and uniform sand piles are calculated individually at first. Then the bearing capacity of SCP improved ground is calculated by the weighted average of these bearing capacities and incorporating the replacement area ratio, as, (see Figure 2.6):
   Bearing capacity:

$$\begin{aligned} P &= q_a \cdot A \\ &= \{as \cdot q_{as} + (1 - as) \cdot q_{ac}\} \cdot A \end{aligned} \qquad (2.12)$$

bearing capacity of clay ground:

$$q_{ac} = \frac{1}{Fs} c \cdot N_c \qquad (2.12a)$$

bearing capacity of sandy ground:

$$q_{as} = \frac{1}{Fs} \cdot \frac{1}{2} \cdot B \cdot \gamma_s \cdot N_\gamma \qquad (2.12b)$$

where:
   as : replacement area ratio
   A : cross sectional area of foundation (m^2)
   B : width of foundation (m)
   c : shear strength of clay ground (kN/m^2)

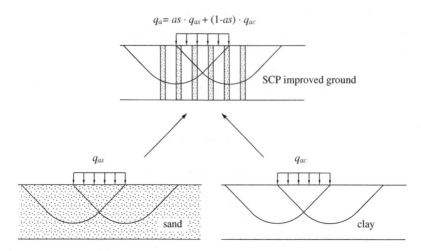

Figure 2.6.   Bearing capacity calculation by Terzaghi's theory.

Fs : safety factor
N_c : bearing capacity factor for cohesion
N_γ : bearing capacity factor for self-weight
P : bearing capacity (kN)
q_a : bearing capacity of improved ground (kN/m^2)
q_{ac} : bearing capacity of clay ground (kN/m^2)
q_{as} : bearing capacity of sandy ground having same property as sand piles (kN/m^2)
γ_s : unit weight of sand (kN/m^3)

2.4   STABILITY

2.4.1   Slip circle analysis

In the case of the stability problem of SCP improved ground, the stability is usually evaluated by a slip circle analysis in which shear strength of composite ground is incorporated (see Figures 2.7 and 2.8). This analysis can also be applicable to the bearing capacity calculation of improved ground (see Figure 2.8). The safety factor, Fs, and the bearing capacity, P, can be calculated as Equations (2.13) and (2.14) respectively.
Safety factor:

$$Fs = \frac{M_R}{M_D}$$
$$= \frac{R \cdot \Sigma(\tau \cdot \Delta l)}{\Sigma(W \cdot x)} \tag{2.13}$$

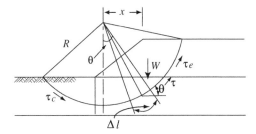

Figure 2.7.   Slip circle analysis for slope stability.

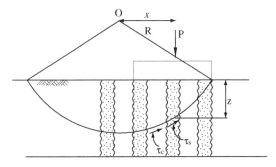

Figure 2.8.   Slip circle analysis for bearing capacity.

Bearing capacity:

$$P = \frac{1}{Fs} \cdot \frac{R \cdot \Sigma(\tau \cdot \Delta l)}{x}$$ (2.14)

where:
  M_D : driving moment (kNm)
  M_R : resisting moment (kNm)
  R : radius of slip circle (m)
  W : weight of embankment (kN)
  x : horizontal distance of weight of embankment or external load measured from center of slip circle (m)
  Δl : arc of slip circle (m)
  τ : shear strength of improved ground (kN/m^2)

　　Two slip circle analyses have been proposed for evaluating the stability of clay and sandy ground: the modified Fellenius method and the Bishop method. Among them, the modified Fellenius method has been frequently adopted for analysis of SCP improved ground. The modified Fellenius method assumes that the direction of resultant force acting on vertical planes between the segments is parallel to the base of the segments (slice method). Since the vertical external load is distributed toward only the vertical

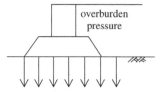

Figure 2.9a.   Slice method.

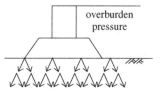

Figure 2.9b.   Stress distribution method.

direction in the slice method as shown in Figure 2.9a, vertical stress increment occurs in the area only beneath the superstructure and external loading area. As a result, additional shear strength increment does not mobilize in the sand piles installed in the periphery of the superstructure and/or loading area, but the vertical stress due to superstructure and/or external load acts on the sand piles only beneath them. In the case of the slice method, the sand piles installed in the periphery of the superstructure and/or external loading area are not subjected to its overburden pressure and do not mobilize shear strength so much.

The stress distribution method, on the other hand, incorporates the vertical stress distribution in the ground due to superstructure and/or external load as shown in Figure 2.9b. Therefore the sand piles installed in the periphery of the superstructure and/or external loading area is also subjected to a part of the overburden pressure(s). This indicates that the shear strength of the sand piles not only beneath the super-structure but also in the periphery are incorporated to evaluate the stability. It is found from calculations that the modified Fellenius method with the stress distribution method provides a higher safety factor than the modified Fellenius method with the slice method as shown in Figure 2.10 [3]. This means that a more economical design can be obtained by the stress distribution method.

The stress distribution in SCP improved ground is usually estimated by Boussinesq's equations with the assumption that it has uniform properties, as shown in Figures 2.11a and 2.11b and Equations (2.15a) and (2.15b) for uniform load and for triangular load respectively (see Figures 2.11a and 2.11b).

For uniform load:

$$\Delta\sigma_z = \frac{p}{2\pi} \{2(\alpha_2 - \alpha_1) - \sin 2\alpha_1 + \sin 2\alpha_2\} \tag{2.15a}$$

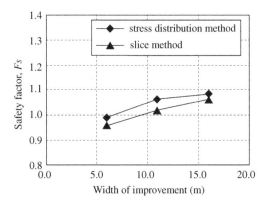

Figure 2.10.   Comparison of safety factor calculated by the stress distribution method and the slice method [3].

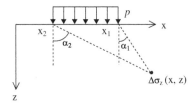

Figure 2.11a.   Vertical stress distribution in ground for uniform load.

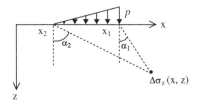

Figure 2.11b.   Vertical stress distribution in ground for triangular load.

For triangular load:

$$\Delta\sigma_z = \frac{p}{\pi}\left\{\frac{(\alpha_1 - \alpha_2) \cdot \cos\alpha_1 \cdot \cos\alpha_2}{\sin(\alpha_1 - \alpha_2)} - \cos\alpha_1 \cdot \sin\alpha_1\right\} \qquad (2.15b)$$

where:

p : external load (kN/m^2)

$\Delta\sigma_z$: increment of vertical stress in ground due to external load (kN/m^2)

$$\alpha_1 = \tan^{-1}\left(\frac{x - x_1}{z}\right)$$

$$\alpha_2 = \tan^{-1}\left(\frac{x - x_2}{z}\right)$$

2.4.2   Shear strength formula

Four formulas (Equations (2.16a) to (2.16d)) for evaluating the shear strength of composite ground for the slip circle analysis have been proposed [4–5] (see Figure 2.12), which are summarized as follows:

Formula (1):

$$\tau = (1 - as) \cdot (c_0 + kz + \mu_c \cdot \Delta\sigma_z \cdot c_u/p \cdot U)$$
$$+ (\gamma_s \cdot z + \mu_s \cdot \Delta\sigma_z) \cdot as \cdot \tan\phi_s \cdot \cos^2\theta \tag{2.16a}$$

Formula (2):

$$\tau = (1 - as) \cdot (c_0 + kz)$$
$$+ (\gamma_m \cdot z + \Delta\sigma_z) \cdot \mu_s \cdot as \cdot \tan\phi_s \cdot \cos^2\theta \tag{2.16b}$$

Formula (3):

$$\tau = (\gamma_m \cdot z + \Delta\sigma_z) \cdot \tan\phi \cdot \cos^2\theta \tag{2.16c}$$

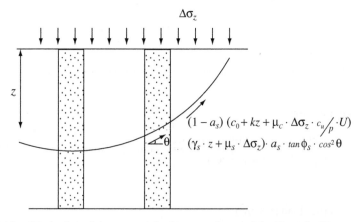

Figure 2.12.   Illustration of shear strength of improved ground for Formula (1).

Formula (4):

$$\tau = (\gamma_m \cdot z + \Delta\sigma_z) \cdot \tan\phi_m \cdot \cos^2\theta \qquad\qquad (2.16d)$$

where:

as : replacement area ratio
c_0 : shear strength of clay at ground surface (kN/m^2)
c_u/p : shear strength increment ratio
k : increment ratio of shear strength of clay with depth (kN/m^3)
n : stress concentration ratio

$$n = \frac{\sigma_s}{\sigma_c}$$

U : average degree of consolidation
z : depth (m)
$\Delta\sigma_z$: increment of vertical load intensity (kN/m^2)
γ_s : unit weight of sand pile (kN/m^3)
γ_m : average unit weight of improved ground (kN/m^3)

$$\gamma_m = \gamma_s \cdot as + \gamma_c \cdot (1 - as)$$

θ : inclination of slip circle
μ_c : stress concentration coefficient of clay ground for external load

$$\mu_c = \frac{\sigma_c}{\sigma} = \frac{1}{1 + (n-1) \cdot as}$$

μ_s : stress concentration coefficient of sand pile for external load

$$\mu_s = \frac{\sigma_s}{\sigma} = \frac{n}{1 + (n-1) \cdot as}$$

σ_c : vertical stress on clay ground (kN/m^2)
σ_s : vertical stress on sand pile (kN/m^2)
τ : shear strength of improved ground (kN/m^2)
ϕ : internal friction angle of sand
ϕ_m : internal friction angle of sand pile

$$\phi_m = \tan^{-1}(\mu_s \cdot as \cdot \tan\phi_s)$$

ϕ_s : internal friction angle of sand pile

Formula (1)

Formula (1) (Equation (2.16a)) was derived by the design concept proposed by Murayama [1] in principle and has been modified through many design experiences. The principle of the formula is derived by summing up the shear strengths mobilized along sand piles and clay ground individually; the first term of the equation represents the undrained shear strength mobilized along clay ground in which the shear strength increase due to external load(s) is also incorporated. The second term represents the shear strength mobilized along sand piles. This formula assumes that the ultimate shear strength is mobilized along sand piles and clay ground simultaneously. However, the well-known phenomenon that stress-strain relationships of sand and clay are quite different might suggest that this assumption is not correct. Triaxial compression tests on clay specimens improved by a sand pile have revealed that the shear strength of SCP improved ground could be evaluated by summing up the shear strengths of sand piles and clay ground [6].

Two parameters, μ_c and μ_s, are introduced to incorporate the stress concentration effect. The vertical stress at a depth, z, is estimated as Equation (2.17) for sand pile and clay ground respectively. The formula indicates that the sand pile and the surrounding clay support their self-weight by themselves, but support external load(s) according to the stress concentration ratio of sand pile and surrounding clay ground.

This formula has mainly been adopted for on-land construction and marine construction as a design code in Japan, e.g. [4–5]. As the shear strength formulation of SCP improved ground contains the stress concentration effect, few ordinary slip circle programs can be used for the calculation. A slip circle analysis combining this shear strength formulation and the stress distribution method was provided by Port and Airport Research Institute and Fudo Construction Co., Ltd. in 2003 [7]. This formulation is very important in the case of step-wise loading and/or SCP improved grounds with a low replacement area ratio where the shear strength of clay between sand piles becomes an important rule for stability.

$$
\begin{aligned}
\sigma_{z,sandpile} &= \gamma_s \cdot z + \mu_s \cdot \Delta\sigma_z \\
\sigma_{z,clayground} &= \gamma_c \cdot z + \mu_c \cdot \Delta\sigma_z
\end{aligned}
\qquad (2.17)
$$

where:
z : depth (m)
σ_z : vertical stress at a depth, z (kN/m^2)
γ_c : unit weight of clay ground (kN/m^3)
γ_s : unit weight of sand pile (kN/m^3)
μ_c : stress concentration coefficient of clay ground for external load
μ_s : stress concentration coefficient of sand pile for external load
$\Delta\sigma_z$: increment of vertical stress due to external load (kN/m^2)

Formula (2)

Formula (2) (Equation (2.16b)), proposed by Tamura (1974), is similar to Equation (2.16a) in which the shear strength is derived by summing up the shear strengths mobilized along sand piles and clay ground individually. In the first term of the equation, no shear strength increase due to external load(s) is incorporated. The formulation of the shear strength of sand piles is slightly different from Formula (1), in which an average unit weight of improved ground, γ_m, is used instead of that of a sand pile, γ_s. Furthermore, the stress concentration coefficient of sand pile, μ_s, covers not only the term of external load but also the term of self weight, $\gamma_m z$. The vertical stress at a depth, z, is estimated as Equation (2.18). This means that the sand piles installation causes the vertical stress distribution to be changed from the initial condition. Therefore the definition of the stress concentration coefficients, μ_c and μ_s, in Formula (2) (Equation (2.16b)) should be different from that in Formula (1) (Equation (2.16a)), but the same μ_c and μ_s values as Formula (1) are usually used in the practical design.

$$\sigma_{z,sandpile} = \mu_s \cdot \gamma_m \cdot z + \mu_s \cdot \Delta\sigma_z$$
$$\sigma_{z,clayground} = \mu_c \cdot \gamma_m \cdot z + \mu_c \cdot \Delta\sigma_z \qquad (2.18)$$

where:

z : depth (m)

σ_z : vertical stress at a depth, z (kN/m^2)

γ_m : unit weight of improved ground (kN/m^3)

μ_c : stress concentration coefficient of clay ground for external load

μ_s : stress concentration coefficient of sand pile for external load

$\Delta\sigma_z$: increment of vertical stress due to external load (kN/m^2)

Formula (3)
Formula (3) (Equation (2.16c)) is formulated by the shear strength of sand piles alone, while the shear strength mobilized along clay ground is not incorporated. This formula is applicable to improved grounds with a high replacement area ratio, generally exceeding 0.7. Formula (3) assumes that a sandy ground with uniform characteristics is constructed. This equation is sometimes used in practical design with an internal friction angle of 'sandy ground', ϕ, of 30 or 35 degrees.

Formula (4)
Formula (4) (Equation (2.16d)) is a typical simplified case of Formula (3), in which the shear strength of sand piles is incorporated alone with an averaged internal friction angle of sand piles, ϕ_m. Similar to Formula (3), this formula is also applicable to improved grounds with a high replacement area ratio exceeding 0.7. The internal friction angle, ϕ_m, of 'sandy ground' is estimated by incorporating the composite ground consisting of sand piles and clay, and is expressed as $\phi_m = \tan^{-1}(\mu_s \cdot as \cdot \tan \phi_s)$.

The magnitude of stress concentration ratio usually used in the practical design is summarized in Tables 2.1 [4–5] and 2.2 [8], which is determined not only by field experiences but also by back calculation of field experiences.

Table 2.1.   Magnitude of stress concentration ratio adopted in shear strength
Formula (1) [4–5].

Replacement area ratio, *as*	Stress concentration ratio, *n*	Internal friction angle of compacted sand pile, ϕ_s
Less than 0.4	3	30
0.4 to 0.7	2	30
Higher than 0.7	1	35

Table 2.2.   Magnitude of stress concentration ratio and internal friction angle adopted in shear strength formula [8].

Shear strength formula	Stress concentration ratio, *n*	Internal friction angle, ϕ_s	Unit weight, γ_s' (kN/m³)	Safety factor, *Fs*	Remarks
(1)	3 to 5	30 to 35	9.8	1.2 to 1.3	applicable for all *as*
(2)	1	30 to 33.4	9.8	>1.1	*as* > 0.3
	2			1.2	
(3)	–	–	–	1.3	*as* ⩾ 0.7
(4)	1	34 to 35	9.8	1.25 to 1.3	*as* ⩾ 0.7
	2	30			

* internal friction angles in the table are estimated by the SPT N-value according to Dunham's equation, $\phi_s = \sqrt{12 \cdot N} + 20$, where ϕ_s = 33.4 and 34 for SPT N-value = 15 and 16 respectively.

2.5   SETTLEMENT

In this section, the settlement of fixed type SCP improved ground is briefly intro-duced. The settlement of floating type SCP improved ground will be described in Chapter 8.

2.5.1   Amount of settlement

In the current design procedure in Japan, the consolidation settlement of fixed type SCP improved ground is usually calculated while undrained settlement of SCP improved ground is seldom considered [4–5]. In the settlement calculation, it is usu-ally assumed that compacted sand piles and surrounding clay ground settle uniformly as illustrated in Figure 2.13. The stress concentration effect is also incorporated. This assumption is applicable to flexible loading such as embankment, because external load acting on the ground surface can be assumed to concentrate rapidly on sand piles at very shallow depth [9]. The final consolidation settlement of improved ground, S, is

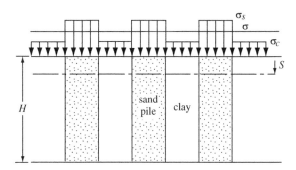

Figure 2.13.   Illustration of uniform settlement of improved ground.

calculated by multiplying the final consolidation settlement of the original ground without any improvement, S_0, and a settlement reduction factor, β, as formulated by Equation (2.19). This formulation is derived by an assumption of one-dimensional deformation of sand piles and clay ground. The settlement reduction factor, β, is derived by the stress concentration effect on sand piles. The final consolidation settlement of original ground is often calculated by Terzaghi's theory as expressed in Equations (2.20a) to (2.20c), which has been widely adopted for many types of constructions. In the case of a multiple-layer system, the settlement should be calculated by summing up the settlements taking place in each layer.

$$S = \beta \cdot S_0$$
$$\beta = \frac{1}{1 + (n - 1) \cdot as} \tag{2.19}$$

$$S_0 = \frac{\Delta e}{1 + e_0} H \tag{2.20a}$$

$$S_0 = m_v \cdot \sigma \cdot H \tag{2.20b}$$

$$S_0 = H \cdot Cc \cdot \log \frac{\sigma}{\sigma_0} \tag{2.20c}$$

where:
Cc : compressive coefficient of clay
e_0 : initial void ratio of clay ground
H : ground thickness (m)
m_v : volumetric coefficient of clay (m²/kN)
S : settlement of improved ground (m)
S_0 : settlement of original (unimproved) ground (m)
β : settlement reduction factor

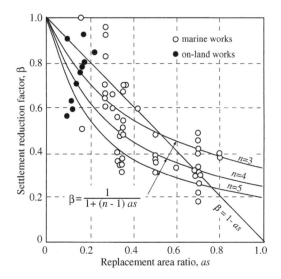

Figure 2.14.   Relationship between replacement area ratio and settlement reduction factor [10].

σ : vertical stress (kN/m²)
σ_0 : initial vertical stress (kN/m²)
Δe : increment of void ratio of clay ground

Figure 2.14 shows the relationship between replacement area ratio, as, and settlement reduction factor, β [10]. The full lines in the figure are derived by Equation (2.19). The settlement reduction factor, β, decreases with increasing replacement area ratio irrespective of the stress concentration ratio. The field data back calculated by measured or estimated settlements of improved ground and unimproved ground are also plotted. Although there is much scatter in the field data, the field data are comparatively close to the formulation with a stress concentration ratio of 3 to 5. Beside this, the settlement reduction factor, $\beta = 1 - as$, has been frequently adopted in the practical design for the case of the replacement area ratio, as, exceeding 0.7.

2.5.2   Speed of settlement

The consolidation phenomenon of SCP improved ground is influenced by many factors such as stiffness and dilatancy characteristics of sand piles, soil disturbance of surrounding clay, initial stress condition, etc., which are very different from the theoretical assumptions. The phenomenon has not yet been clarified even though many studies have been conducted as described later in Chapter 7. According to this, compacted sand piles are simply assumed to function as a drainage path without any stress concentration effect in the practical design procedures, but a modified factor is introduced to the

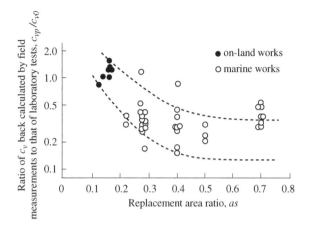

Figure 2.15.   Relationship between the ratio of coefficient of consolidation and replacement area ratio [8].

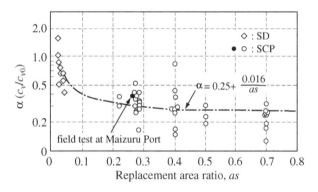

Figure 2.16.   Relationship between the ratio of coefficient of consolidation and replacement area ratio [11].

coefficient of consolidation [4–5]. It is found in laboratory tests and theoretical investigations that the consolidation phenomenon proceeds faster in SCP improved ground than that estimated by Barron's theory. However, the accumulated field data have revealed a quite opposite phenomenon that the consolidation speed of SCP improved ground was somewhat delayed from that estimated by Barron's theory.

Figures 2.15 and 2.16 show the relationship between the ratio of coefficient of consolidation, α, and replacement area ratio, as [8,11]. The ratio of coefficient of consolidation, α, is defined as the ratio of coefficient of consolidation of SCP improved ground back calculated by field measurements, C_{vp}, to that measured in the laboratory oedometer test, C_{v0}. Although there is much scatter in the data, the figure clearly shows that the ratio is quite small and decreases with increasing replacement area ratio, which indicates that consolidation is delayed at relatively high replacement area ratios. The reason for the delay in consolidation is considered to be the soil disturbance effect during sand piles installation, where the coefficient of consolidation of disturbed clay is much smaller than that of undisturbed clay. The figure also indicates another interesting phenomenon at a low replacement area ratio, in which the consolidation of improved ground proceeds faster than the estimation. This is considered to be due to the stress concentration effect where the stress acting on clay ground decreases as the consolidation proceeds.

2.6   EARTH PRESSURE

2.6.1   Active and passive earth pressures

The SCP improvement is also used to decrease active earth pressure and/or increase passive earth pressure. The property of SCP improved ground consisting of sand piles and surrounding clay has not been investigated in detail and is not well understood. Beside this, there are three kinds of earth pressure formula, Equations (2.21) to (2.23), for evaluating the static and dynamic earth pressures of SCP improved ground with infinite width (see Figures 2.17a and 2.17b) [8]. Formulas (1) and (3) are usually adopted for improved grounds with a medium replacement area ratio of 0.4 to 0.7, while Formula (2) is usually adopted for improved grounds with a high replacement area ratio exceeding 0.7. None of them incorporates the effect of geometric arrangement of sand piles in a ground. As it is not thoroughly clarified yet, a safe-side design is recommended in practical designs in which the replacement area ratio should exceed 0.4 and the stress concentration ratio should be unity.

Formula (1):

$$p_A = (1 - as) \cdot p_{CA} + as \cdot p_{SA} \tag{2.21a}$$

$$p_P = (1 - as) \cdot p_{CP} + as \cdot p_{SP} \tag{2.21b}$$

where:
    as : replacement area ratio
    p_A : active earth pressure of improved ground (kN/m^2)
    p_P : passive earth pressure of improved ground (kN/m^2)
    p_{CA} : active earth pressure of uniform clay ground (kN/m^2)

$$p_{CA} = \sigma - 2c_u$$

    p_{CP} : passive earth pressure of uniform clay ground (kN/m^2)

$$p_{CP} = \sigma + 2c_u$$

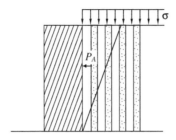

Figure 2.17a.   Illustration of active earth pressure of improved ground.

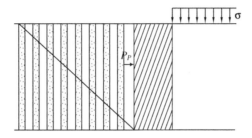

Figure 2.17b.   Illustration of passive earth pressure of improved ground.

p_{SA} : active earth pressure of uniform sandy ground (kN/m²)

$$p_{SA} = (\sigma + \gamma_s \cdot z) \cdot \tan^2 \left(\frac{\pi}{4} - \frac{\phi_s}{2} \right)$$

P_{SP} : passive earth pressure of uniform sandy ground (kN/m²)

$$p_{SP} = (\sigma + \gamma_s \cdot z) \cdot \tan^2 \left(\frac{\pi}{4} + \frac{\phi_s}{2} \right)$$

z : depth (m)
γ_s : unit weight of sand pile (kN/m³)
ϕ_s : internal friction angle of sand pile
σ : overburden pressure (kN/m²)

Formula (2):

$$p_A = (\sigma + \gamma_m \cdot z) \cdot \tan^2 \left(\frac{\pi}{4} - \frac{\phi_m}{2} \right) \qquad\qquad (2.22a)$$

$$p_P = (\sigma + \gamma_m \cdot z) \cdot \tan^2 \left(\frac{\pi}{4} + \frac{\phi_m}{2} \right) \qquad\qquad (2.22b)$$

where:

    as : replacement area ratio
    p_A : active earth pressure of improved ground (kN/m²)
    p_P : passive earth pressure of improved ground (kN/m²)
    z : depth (m)
    γ_m : unit weight of improved ground (kN/m³)

$$\gamma_m = as \cdot \gamma_s + (1 - as) \cdot \gamma_c$$

    ϕ_m : internal friction angle of improved ground

$$\phi_m = \tan^{-1}(as \cdot \tan \phi_s)$$

    σ : overburden pressure (kN/m²)

Formula (3):

$$p_A = K_A \cdot (\gamma_m \cdot z + \sigma) - 2 \cdot (1 - as) \cdot c \cdot \sqrt{K_A}$$

$$K_A = \frac{\cos^2(\phi_m - \theta_0 - \theta)}{\cos\theta_0 \cdot \cos^2\theta \cdot \cos(\theta + \theta_0 + \delta)\left[1 + \sqrt{\dfrac{\sin(\phi_m + \delta) \cdot \sin(\phi_m - \alpha - \theta_0)}{\cos(\theta + \theta_0 + \delta) \cdot \cos(\theta - \alpha)}}\right]^2}$$

(2.23a)

$$p_P = K_P \cdot (\gamma_m \cdot z + \sigma) + 2 \cdot (1 - as) \cdot c \cdot \sqrt{K_P}$$

$$K_P = \frac{\cos^2(\phi_m - \theta_0 + \theta)}{\cos\theta_0 \cdot \cos^2\theta \cdot \cos(\theta - \theta_0 + \delta)\left[1 - \sqrt{\dfrac{\sin(\phi_m - \delta) \cdot \sin(\phi_m + \alpha - \theta_0)}{\cos(\theta - \theta_0 + \delta) \cdot \cos(\theta - \alpha)}}\right]^2}$$

(2.23b)

where:

    k : coefficient of seismic intensity
    p_A : active earth pressure of improved ground (kN/m²)
    p_P : passive earth pressure of improved ground (kN/m²)
    z : depth (m)
    α : inclination angle of ground surface (see Figure 2.18)
    γ_m : unit weight of improved ground (kN/m³)

$$\gamma_m = as \cdot \gamma_s$$

    δ : friction angle of earth pressure on retaining wall (see Figure 2.18)
    θ : inclination angle of retaining wall (see Figure 2.18)

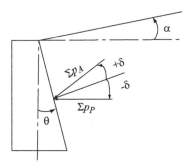

Figure 2.18.   Illustration of earth pressure acting on retaining wall.

θ_0 : angle of seismic intensity

$$\theta_0 = \tan^{-1} k_h$$

ϕ_m : internal friction angle of improved ground

Formula (1)
In this formula, the earth pressures of uniform clay ground and uniform sandy ground are calculated individually by the conventional method at first, e.g. Rankin's theory. Then the earth pressure of SCP improved ground is calculated by the weighted average of these pressures with respect to the replacement area ratio, *as*. This formula has usually been adopted for improved grounds with a replacement area ratio ranging from 0.4 to 0.7 [8].

Formula (2)
In this formula, the earth pressure of improved ground is assumed to be that of sandy ground with uniform properties. The unit weight and internal friction angle, γ_m and ϕ_m, are estimated by incorporating the replacement area ratio, *as*. It should be noted that the definition of internal friction angle, ϕ_m, is different from that in the bearing capacity calculation (see Chapter 2.4.2). The reason for the difference is thought to be that the horizontal earth pressure might not be influenced so much by the vertical stress distribution. This method has usually been adopted for improved grounds with a relatively high replacement area ratio exceeding 0.7, where sand piles are almost in contact with each other.

Formula (3)
In this formula, the earth pressure of improved ground is calculated by assuming uniform ground having cohesion and internal friction angle, which is the same equation as proposed by Mizuno *et al.* [12]. The cohesion, internal friction angle and unit weight of improved ground are estimated by incorporating the replacement area ratio, *as*. However, this formula does not incorporate the effect of geometric arrangement of

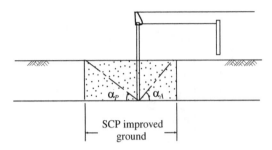

Figure 2.19.   Extent of improved ground criteria (1).

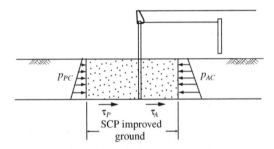

Figure 2.20.   Extent of improved ground criteria (2).

sand piles in a ground. The effects of inclined wall and ground surface are incorporated to estimate the earth pressure as well as the effect of inertia load.

2.6.2   Extent of improved ground

As the number of researches is rather limited (e.g. [13]), the effect of the extent of SCP improved ground on earth pressure has not yet been clarified. Here, a proposal on the minimum extent (depth and width) of SCP improved ground is introduced, which gives three criteria [8].

(1) The minimum extent of improved ground should be the portion containing the active and passive failure lines as shown in Figure 2.19. The inclinations of the active and passive failure lines, α_A and α_P, are usually assumed as about 45 and 30 degrees respectively.
(2) The minimum extent of improved ground should be the portion where the active and passive earth pressures are satisfied, as given by Equations (2.24a) and (2.24b) respectively (see Figure 2.20).

for active earth pressure:

$$P_A \geq P_{AC} - \tau_A \qquad (2.24a)$$

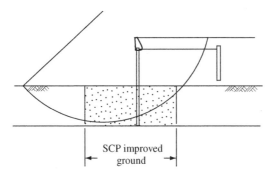

Figure 2.21.   Extent of improved ground criteria (3).

for passive earth pressure:

$$P_P \leq P_{PC} + \tau_P \tag{2.24b}$$

where:

p_A : active earth pressure of improved ground (kN/m²)
p_{AC} : active earth pressure of unimproved ground (kN/m²)
p_P : passive earth pressure of improved ground (kN/m²)
p_{PC} : passive earth pressure of unimproved ground (kN/m²)
τ_A : shear strength on bottom of improved ground at active side (kN/m²)
τ_P : shear strength on bottom of improved ground at passive side (kN/m²)

(3) The minimum extent of improved ground should be the portion through which the slip circle passes with a safety factor of higher than 1.2 or 1.3 (see Figure 2.21).

2.7   HORIZONTAL RESISTANCE

2.7.1   Basic concept

The SCP method has also been used to improve the horizontal resistance of a pile, pile group and sheet wall. The horizontal resistance of these structures is usually evaluated by the ground subgrade reaction theory as shown in Equation (2.25) and Figure 2.22 irrespective of ground condition and type of ground improvement. In the analysis, the improved ground is assumed to be an elastic material having a coefficient of subgrade reaction, k_h:

$$EI \frac{d^4 y}{dz^4} = -P = -p \cdot B \tag{2.25}$$

where:

B : diameter or width of pile structure (m)
EI : flexural rigidity of pile structure (kN m²)

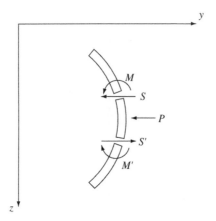

Figure 2.22.   Illustration of subgrade reaction theory.

p : subgrade reaction pressure (kN/m^2)
P : subgrade reaction force (kN)
y : horizontal displacement of pile structure (m)
z : depth (m)

The subgrade reaction pressure can be expressed as the following equation:

$$p = k_h \cdot z^m \cdot y^n \tag{2.26}$$

where:
k_h : coefficient of subgrade reaction (kN/m^3)
m : coefficient (either 0 or 1)
n : coefficient (either 1 or 0.5 in general)
p : subgrade reaction pressure (kN/m^2)

The parameter, n, is selected as either 1 or 0.5, according to whether the ground property is assumed to be linear or nonlinear elastic, respectively. The equation becomes the same equation as Chang's method when n is 1, and the same as the Port and Harbour Research Institute Method (PHRI Method) when n is 0.5 [4–5]. The parameter m defines the inclination of coefficient of subgrade reaction with depth, and is usually considered to be either 0 (uniform coefficient through depth) or 1 (linear increase with depth). In PHRI Method, m of 1 is usually adopted for sandy ground, because the coefficient of subgrade reaction increases with increasing overburden pressure. For clay grounds, the parameter m is usually considered to be 0 for grounds having uniform undrained shear strength distribution with depth and to be 1 for grounds having undrained shear strength increasing with depth. The coefficient of subgrade reaction has been investigated thoroughly for clay and sandy grounds.

2.7.2   Formulation of coefficient of subgrade reaction

Many research efforts have been performed experimentally and analytically on the horizontal resistance of clay, sandy and SCP improved grounds. Three kinds of formulas for the coefficient of subgrade reaction have been proposed for SCP improved ground and are adopted in current designs as shown in Equations (2.27a) to (2.27c) [14].

Formula (1):

$$k_h = as \cdot k_{hs} + (1 - as) \cdot k_{hc} \qquad\qquad (2.27a)$$

Formula (2):

$$\begin{aligned} k_h &= k_{h0} \cdot D^{-1/4} \cdot y^{-1/2} \\ k_{h0} &= as \cdot k_{hs0} + (1 - as) \cdot k_{hc0} \end{aligned} \qquad\qquad (2.27b)$$

Formula (3):

$$k_h = 0.15 \cdot N \qquad\qquad (2.27c)$$

where:
  as : replacement area ratio
  D : diameter of pile structure (cm)
  k_h : coefficient of subgrade reaction of SCP improved ground (kN/m^3)
  k_{hs} : coefficient of subgrade reaction of sand pile (kN/m^3)
  k_{hc} : coefficient of subgrade reaction of original ground (kN/m^3)
  k_{h0} : coefficient of subgrade reaction of SCP improved ground at 1 cm horizontal displacement of 1 cm diameter pile (kN/m^3)
  k_{hc0} : coefficient of subgrade reaction of clay ground at 1 cm horizontal displacement of 1 cm diameter pile (kN/m^3)
  k_{hs0} : coefficient of subgrade reaction of sandy ground at 1 cm horizontal displacement of 1 cm diameter pile (kN/m^3)

$$k_{h0} = as \cdot k_{hs0} + (1 - as) \cdot k_{hc0}$$

$$k_{hs0} = 64.68 \quad (MN/m^3) \approx 66000 \quad (kN/m^3)$$

$$k_{hc0} = 5.11 \cdot q_u^{0.934} \approx 5100 \cdot q_u \quad (kN/m^3)$$

  N : SPT N-value of improved ground
  q_u : unconfined compressive strength of clay ground (kN/m^2)
  y : horizontal displacement of pile structure (cm)

As the unit system is not consistent in the above formulas and the constants in these formulas have various units, it should be noted that these formulations are applicable to specific combinations of units for each parameter.

Formula (1)

Formula (1) (Equation (2.27a)) is based on a similar concept as Formula (1) for the earth pressure evaluation (Equation (2.21a)). In this method, the coefficient of subgrade reaction of SCP improved ground is calculated based on the series connection of springs and by weighted average of those of compacted sand piles and original clay ground. The coefficient of compacted sand piles is usually estimated assuming the SPT N-value of 10 to 15 and the accumulated field data as shown in Figure 2.23 [15]. This formula is often adopted for improved grounds whose replacement area ratio exceeds around 0.4.

Figure 2.23 shows the relationship between the SPT N-value and coefficient of subgrade reaction [15]. The data shown in the figure are the coefficient at a pile displacement of 1 cm, which were estimated by back calculating the load displacement curves in field loading tests. Yokoyama [15] pointed out that the value of the coefficient had relatively large dispersion of around 100%, and the SPT N-value to be adopted in the figure should be an average value measured from the ground surface to the depth of characteristic length, β^{-1}, which is defined as Equation (2.28). This scatter in data does not greatly influence the magnitude of maximum bending

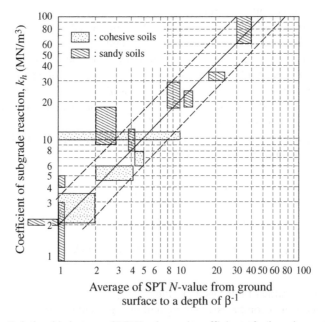

Figure 2.23.   Relationship between SPT N-value and coefficient of subgrade reaction [15].

moment of pile structure, but influences the amount of pile deformation more sensitively [16].

$$\beta = \sqrt[4]{k_h \cdot B / 4 \cdot EI} \tag{2.28}$$

where:

B : width of pile (m)
EI : flexural rigidity of pile (kN/m^2)
k_h : average of coefficient of subgrade reaction (kN/m^3)
$1/\beta$: characteristic length (m)

Figure 2.24 shows the relationship between the SPT N-value and coefficient of subgrade reaction, which is used for the PHRI Method for S type and C type grounds [4–5]. The S type and C type grounds are classified by whether the coefficient of subgrade reaction increases linearly with depth (S type) or is uniform through depth (C type). In general, sandy grounds and normally consolidated clay grounds are categorized as S type ground, while clay ground with uniform shear strength is categorized as C type ground. The coefficient of subgrade reaction is usually expressed as k_{hs} and k_{hc} for S type and C type grounds in the PHRI Method. As shown in Equation (2.26), the k_{hs} and k_{hc} in the PHRI Method have the unit of kN/m$^{3.5}$ and kN/m$^{2.5}$ respectively. In Figure 2.24a, the average increment of SPT N-value per meter is plotted on the horizontal axis. The figure shows that the coefficient increases linearly with increasing

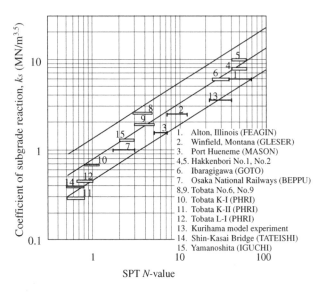

1. Alton, Illinois (FEAGIN)
2. Winfield, Montana (GLESER)
3. Port Hueneme (MASON)
4,5. Hakkenbori No.1, No.2
6. Ibaragigawa (GOTO)
7. Osaka National Railways (BEPPU)
8,9. Tobata No.6, No.9
10. Tobata K-I (PHRI)
11. Tobata K-II (PHRI)
12. Tobata L-I (PHRI)
13. Kurihama model experiment
14. Shin-Kasai Bridge (TATEISHI)
15. Yamanoshita (IGUCHI)

Figure 2.24a.   Relationship between SPT N-value and coefficient of subgrade reaction for S type ground [4–5].

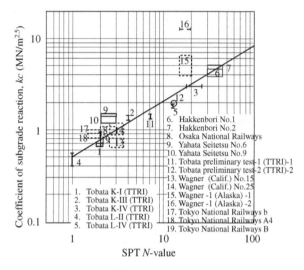

Figure 2.24b.   Relationship between SPT *N*-value and coefficient of subgrade reaction for C type ground [4–5].

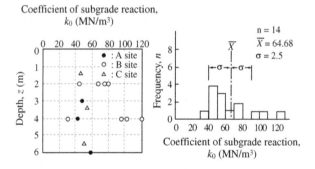

Figure 2.25.   Distribution of coefficient of subgrade reaction with depth [8].

average increment of SPT *N*-value in a double logarithmic scale. Figure 2.24b shows the k_{hc} value against SPT *N*-value. The figure also shows that the coefficient of subgrade reaction increases linearly with increasing SPT *N*-value in a double logarithmic scale.

Formula (2)

The basic concept of Formula (2) is the same as Formula (1) in which the subgrade reaction of improved ground is calculated by the weighted average of those of sandy and clay grounds. The coefficient of subgrade reaction of sandy ground is derived by the field data measured by borehole horizontal loading tests, as shown in Figure 2.25 [8]. These

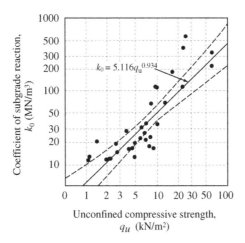

Figure 2.26.   Relationship between coefficient of subgrade reaction and unconfined compressive strength [17].

data in the figure were obtained in compacted sand piles installed in three kinds of clay ground. According to the figure, the coefficient of 64.68 MN/m³ was adopted in the practical design.

The coefficient of subgrade reaction of clay ground is derived by the relationship to unconfined compressive strength as shown in Figure 2.26 [17]. The coefficient in the figure was derived by the field data measured by borehole horizontal loading tests. Nakajima et al. [17] proposed the approximate relationship of $k_0 = 5100 \, q_u$ (kN/m³), in which q_u is unconfined compressive strength. This formula takes into account the effects of the diameter of pile structure and non-linearity of ground by introducing the terms $D^{-1/4}$ and $y^{-1/2}$. This formula is often adopted for SCP improved grounds whose replacement area ratio exceeds around 0.4.

Formula (3)
Formula (3) is adopted for improved grounds with a relatively high replacement area ratio of around 0.7. In the design code for port facilities [4–5], the coefficient of subgrade reaction is obtained by the SPT N-value as Equation (2.29a). A similar relationship is proposed as Equation (2.29b) by Nakajima et al. [17].

$$k_h = 150 \cdot N \qquad (2.29a)$$

$$k_h = 120 \cdot N \qquad (229b)$$

where:
   k_h : coefficient of subgrade reaction (kN/m³)
   N : SPT N-value

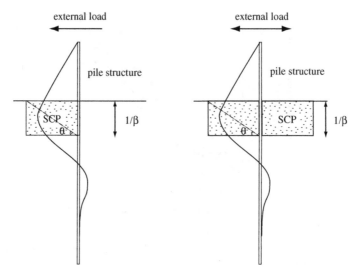

Figure 2.27   (a) Illustration of determination of extent of improved ground for monotonic loading.
(b) Illustration of determination of extent of improved ground for cyclic loading.

2.7.3   Extent of improved ground

The passive (front) side ground of the pile structure should be improved in a case of monotonic horizontal loading, while both sides of the ground should be improved in a case of cyclic loading as shown in Figures 2.27a and 2.27b. Although there have been some researches, the effect of extent of SCP improved ground on the horizontal resistance remains unclear. The minimum extent (depth, D, and width, W) of SCP improved ground is usually obtained as follows (see Figure 2.27):

$$D = 1/\beta \tag{2.30}$$

$$W = 1/\beta \cdot \tan(\theta) \tag{2.31}$$

where:
  D : depth of improved portion (m)
  EI : flexural rigidity of pile (kN/m^2)
  k_h : average of coefficient of subgrade reaction
  W : width of improved portion (m)
  $1/\beta$: characteristic length (m)

$$\beta = \sqrt[4]{(k_h \cdot B)/(4 \cdot EI)}$$

  θ : inclination angle (usually 60 to 70 degree)

REFERENCES

1   Murayama, S.: Considerations of vibro-compozer method for application to cohesive ground. Journal of Construction Machinery, No.150, pp.10–15, 1962 (in Japanese).

2   Sogabe, T.: Technical subjects on design and execution of sand compaction pile method. Proc. of the 36th Annual Conference of the Japan Society of Civil Engineers, III, pp.39–50, 1981 (in Japanese).

3   Coastal Development Institute of Technology: Design case histories of port and harbor facilities, Vol.1, pp.12-1–12-18, 1999 (in Japanese).

4   Ministry of Transport: Technical standards and commentaries for port and harbour facilities in Japan. Ministry of Transport, Japan, 1999 (in Japanese).

5   The Overseas Coastal Area Development Institute of Japan: English version of technical standards and commentaries for port and harbour facilities in Japan. 2002 (in Japanese).

6   Matsuo, M., Kuga, T. and Maekawa, I.: The study on consolidation and shear strength of cohesive soil containing sand pile. Journal of Geotechnical Engineering, Japan Society of Civil Engineers, No.141, pp.42–55, 1967 (in Japanese).

7   http://www.pari.go.jp/bsh/jbn-kzo/jiban/.

8   Japanese Society of Soil Mechanics and Foundation Engineering: Soil improvement methods – survey, design and execution. The Japanese Society of Soil Mechanics and Foundation Engineering, 1988 (in Japanese).

9   Mogami, T., Nakayama, J., Ueda, S., Kuwata, H., Kamata, H. and Taguchi, S.: Laboratory model tests on composite ground (1st Report). Journal of the Japanese Society of Soil Mechanics and Foundation Engineering, 'Tsuchi-to-Kiso', Vol.16, No.8, pp.9–17, 1968 (in Japanese).

10 Ichimoto, E.: Practical design method and calculation of sand compaction pile method. Proc. of the 36th Annual Conference of the Japan Society of Civil Engineers, III, pp.51–55, 1981 (in Japanese).

11 Kanda, K. and Terashi, M.: Practical formula for the composite ground improved by sand compaction pile method. Technical Note of the Port and Harbour Research Institute, No.669, pp.52, 1990 (in Japanese).

12 Mizuno, Y., Yoshida, M. and Tsuboi, H.: Estimation of earth pressure of composite ground. Proc. of the 34th Annual Conference of the Japan Society of Civil Engineers, III, pp.395–396, 1979 (in Japanese).

13 Terashi, M., Kitazume, M. and Kubo, S.: Centrifuge modeling on passive earth pressure of improved ground by sand compaction pile method. Proc. of the 27th Annual Conference of the Japanese Society of Soil Mechanics and Foundation Engineering, pp.2159–2162, 1992 (in Japanese).

14 Japanese Society of Soil Mechanics and Geotechnical Engineering: Prediction and actual behavior of improvement effect. The Japanese Society of Soil Mechanics and Geotechnical Engineering, 2000 (in Japanese).

15 Yokoyama, K.: Design and execution of steel pile. Sankaido Co., Ltd., 1963 (in Japanese).
16 Kikuchi, Y.: Application of SPT N-value in technical standards for port and harbour facilities. Journal of the Foundation Engineering & Equipment, 'Kisoko', Sogodoboku, No.2, pp.31–35, 2003 (in Japanese).
17 Nakajima, H., Itou, T., Takeuchi, T. and Imai, T.: Ground properties at Osaka area by LLT measurements. Journal of Geology, Vol.78, No.4, pp.165–175, 1972 (in Japanese).

Chapter 3

Design Procedures for Sandy Ground

3.1 INTRODUCTION

The main purpose of applying the SCP method to sandy or silty ground is usually to prevent liquefaction or reduce settlement. In order to achieve these purposes, the most effective way is to increase ground density, which usually provides some benefits: (a) strength increase and (b) uniformity increase. In this chapter a current design procedure for sandy ground to prevent liquefaction is described in detail. Again the design procedure introduced in this section are still being modified in accordance with the new findings from many research projects and accumulated experience. Therefore a specific practical design for each site should be produced not only according to the procedure, but also in accordance with ongoing research activities and accumulated know-how.

3.2 LIQUEFACTION MITIGATION

3.2.1 Liquefaction potential assessment

A methodology combining field and laboratory data has been developed and extensively used in the standard design practice in Japan [1–2]. The possibility of liquefaction during an earthquake is usually assessed by the following two steps: (1) particle size distribution and SPT N-value and (2) cyclic triaxial test.

Particle size distribution and SPT N-value
Figure 3.1 shows particle size distributions of soil to determine the possibility of liquefaction [1–3], which was established based on accumulated field experiences. Soils are considered non-liquefiable if their particle size distribution falls outside the 'possibility of liquefaction' zone. Soils with particle size distribution curve falling within the 'high possibility of liquefaction' zone are assumed to be liquefied during an earthquake. For soils with particle size distribution curve falling within the 'possibility of liquefaction' zone, their liquefaction potential is evaluated based on the relationship between equivalent SPT N-value, N_{65}, and equivalent acceleration, a_{eq}, as shown in Figure 3.2.

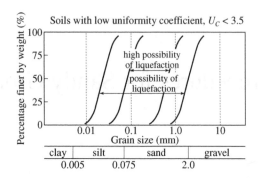

Figure 3.1a. Particle size distribution of soil having possibility of liquefaction for soils with low uniformity coefficient, $U_c < 3.5$ [1–3].

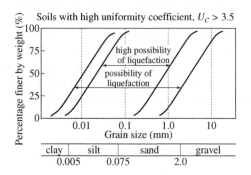

Figure 3.1b. Particle size distribution of soil having possibility of liquefaction for soils with high uniformity coefficient, $U_c > 3.5$ [1–3].

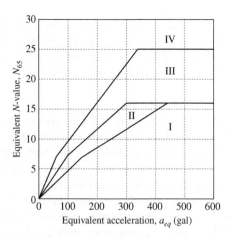

Figure 3.2. Liquefaction potential assessment [1–3].

As the SPT N-value is influenced by an overburden pressure at measured depth, an equivalent SPT N-value, N_{65}, is introduced as Equation (3.1) to incorporate its effect [1–2]. In the equation, the measured SPT N-value is converted to that at an over-burden pressure of 65 kN/m^2, which usually corresponds to the vertical stress at ground water level.

For silty or plastic soils with fines content exceeding 5%, the SPT N-value is also modified to incorporate the effect of fines content. An equivalent acceleration, a_{eq}, on the other hand, is estimated by Equation (3.2).

$$N_{65} = \frac{N_m - 0.019 \cdot \left(\sigma'_v - 65\right)}{0.0041 \cdot \left(\sigma'_v - 65\right) + 1.0} \tag{3.1}$$

where:

N_{65} : equivalent SPT N-value
N_m : measured SPT N-value
σ'_v : effective overburden pressure at measured depth (kN/m^2)

$$a_{eq} = 0.7 \cdot \frac{\tau_{max}}{\sigma'_v} \cdot g \tag{3.2}$$

where:

a_{eq} : equivalent acceleration (gal)
g : acceleration due to gravity (gal)
σ'_v : effective overburden pressure (kN/m^2)
τ_{max} : maximum shear stress (kN/m^2)

According to Figure 3.2, the liquefaction potential can be determined as follows.

i) soils falling within zone I have a very high possibility of liquefaction and are assumed to liquefy during an earthquake.
ii) soils falling within zone II have a high possibility of liquefaction and further eval-uation based on cyclic triaxial test is usually required to determine more precisely.
iii) soils falling within zone III have a low possibility of liquefaction, but further evaluation based on cyclic triaxial test is usually required to determine more precisely.
iv) soils falling within zone IV have a very low possibility of liquefaction and are assumed not to liquefy during an earthquake.

Cyclic triaxial test
For soils falling within zones II or III in Figure 3.2, the liquefaction potential should be evaluated by undrained cyclic triaxial tests on undisturbed soil specimens. A series of cyclic triaxial tests should be carried out to establish the relationship between the

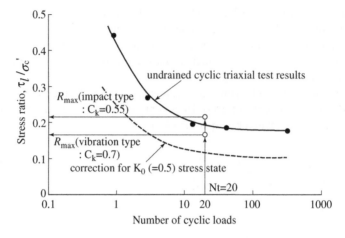

Figure 3.3.   Relationship between the number of cyclic loads and the stress ratio [3].

number of cyclic loads and stress ratio for the soil, as illustrated in Figure 3.3 [3]. The *in-situ* liquefaction resistance, R_{max}, is obtained by Equation (3.3) together with the stress ratio, defined as the stress ratio at 20 cyclic loads.

$$R_{max} = \frac{0.9}{C_k} \cdot \frac{1 + 2K_0}{3} \cdot \left(\frac{\tau_1}{\sigma'_c}\right)_{N=20} \tag{3.3}$$

where:

    C_k : correction factor of input motion (0.55 and 0.7 for impact type and vibration type input motions respectively)

    K_0 : static earth pressure coefficient

    R_{max} : *in-situ* liquefaction resistance

    τ_1/σ'_c : stress ratio

After the seismic stress ratio, L_{max}, is obtained through the site response analysis, the liquefaction potential, F_L, a sort of safety factor against liquefaction, F_L, can be obtained as Equations (3.4) and (3.5). Soils with an F_L value less than unity are assumed to liquefy during an earthquake.

$$L_{max} = \frac{\tau_{max}}{\sigma'_c} \tag{3.4}$$

$$F_L = \frac{R_{max}}{L_{max}} \tag{3.5}$$

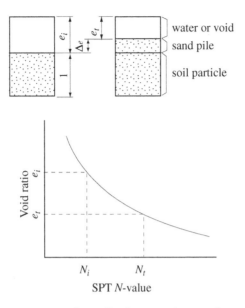

water or void

sand pile

soil particle

Figure 3.4.   Improvement concept for application to sandy ground.

where:

F_L : liquefaction potential

L_{max} : seismic stress ratio

R_{max} : *in-situ* liquefaction resistance

σ'_c : effective consolidation pressure (kN/m^2)

τ_{max} : maximum shear stress obtained from site response analysis (kN/m^2)

3.2.2   Design concept of SCP improvement

When a ground is considered to liquefy during an earthquake according to the above-mentioned procedure, the density of ground should be increased to a certain level to prevent the liquefaction of the ground.

Beside the direct measurement of ground density, the SPT *N*-value has been frequently used as an index to evaluate the property of improved ground. The void ratio of sandy ground is closely related to the SPT *N*-value as illustrated in Figure 3.4. The initial and target void ratios, e_i and e_t, are estimated by measured and target SPT *N*-values respectively. Then, the replacement area ratio, *as*, of improvement for preventing liquefaction is calculated. Then the volume of sand to be installed, *V*, and the arrangement of sand piles are determined according to the replacement area ratio, which will be later shown in Equations (3.7) and (3.8).

Accumulated case records show that the SPT *N*-value is influenced by many factors such as fines content of soil, overburden pressure of ground and horizontal stress in a

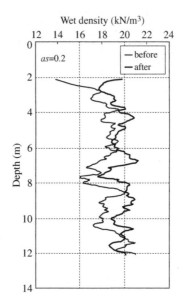

Figure 3.5a.   Influence of ground upheaval on wet density distribution with depth [4].

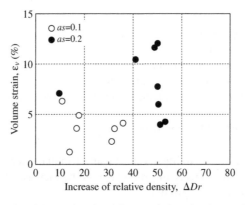

Figure 3.5b.   Influence of ground upheaval on relative density and volumetric strain of ground [4].

ground. These effects should be incorporated to estimate the void ratio more accurately by measured SPT N-values. Beside this, upheaval of ground is usually observed during sand piles installation, particularly in grounds with relatively high fines content. Figure 3.5 shows an example of the influence of ground upheaval on SCP improvement [4]. In Figure 3.5a, the wet density of sandy ground estimated by RI tests is plotted with depth where sand piles are installed with the replacement area ratio of 0.2. Figure 3.5b is the volumetric strain, ε_v, calculated by the measured density increase as shown in Figure 3.5a. It can be seen in the figure that the volumetric strain increases by about 1% to 6% for the replacement area ratio, *as*, of 0.1, and about 4% to 12% for *as* of 0.2, which are smaller than the magnitude of replacement area ratio, *as*. This discrepancy can be explained by the effect of ground upheaval and horizontal deformation of ground during sand piles installation.

As these effects vary greatly for each ground and construction conditions, they have not yet been theoretically clarified. At this moment, there are four practical design procedures for liquefaction prevention as shown in Figure 3.6, which have been established by empirical means together with accumulated field and laboratory measurements. Each design procedure for application to sandy ground is described in detail in the next section.

3.2.3   Design procedure for determination of sand pile arrangement

Figure 3.6 shows the design flow of four procedures for SCP application to sandy ground. Each procedure is explained as follows [5]:

Procedure A
This design procedure is established based on accumulated field data on the relationship between the SPT *N*-value of original ground, replacement area ratio, *as* and SPT *N*-value of improved ground, which are shown in Figures 3.7a and 3.7b [6]. In Figure 3.7a, the SPT *N*-value of sand between compacted sand piles, N_{ti}, is plotted against that of the original ground, N_i, while the SPT *N*-value at the center of compacted sand piles, N_{ts}, is plotted in Figure 3.7b. These figures were obtained by accumulated field data for sandy grounds whose fines content, *Fc*, was less than 20%.

In the design procedure, the magnitude of *Fv* value, which is defined in the same way as *as*, is obtained for the original and target SPT *N*-values according to the figures. As the design is usually conducted with priority on safety, the target SPT *N*-value should be achieved between compacted sand piles where the least improvement effect is achieved. Therefore Figure 3.7a is usually used as a design chart. However, in the case where the average SPT *N*-value of sand piles and surrounding soil is set as the target SPT *N*-value, which is calculated by Equation (3.6), the magnitude of *Fv* value is obtained by iterative calculations with the data in Figures 3.7a and 3.7b.

$$N = (1 - as) \cdot N_{ti} + as \cdot N_{ts} \tag{3.6}$$

where:
    as : replacement area ratio
    N : average SPT *N*-value after improvement
    N_{ti} : SPT *N*-value at sandy ground between sand piles
    N_{ts} : SPT *N*-value at center of sand pile

After obtaining the *Fv* value to achieve the target SPT *N*-value, the volume of sand to be installed per unit depth, *V*, and the diameter of sand piles are obtained by

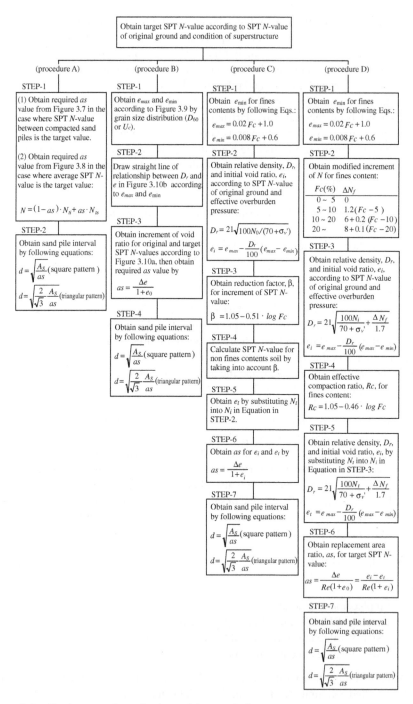

Figure 3.6.   Design procedures for determining sand pile arrangement.

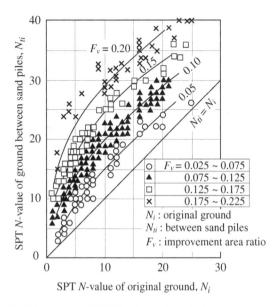

Figure 3.7a.   Relationship between SPT N-values of original and improved ground between sand piles [6].

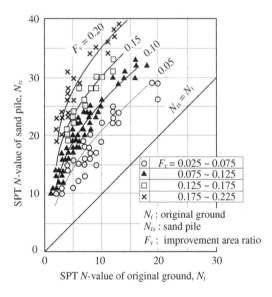

Figure 3.7b.   Relationship between SPT N-values of original and improved ground at center of sand pile [6].

Equations (3.7) and (3.8) for square and equilateral triangular improvement patterns. Here, the Fv value is substituted into the replacement area ratio, as, in the equation.

for square pattern:

$$V = as \cdot D^2 \tag{3.7a}$$

$$d = 2\sqrt{\frac{V}{\pi}}$$
$$= 2 \cdot D\sqrt{\frac{as}{\pi}} \tag{3.7b}$$

for equilateral triangular pattern:

$$V = as \cdot \frac{\sqrt{3}}{2} \cdot D^2 \tag{3.8a}$$

$$d = 2\sqrt{\frac{V}{\pi}}$$
$$= 2 \cdot D\sqrt{\frac{as \cdot \sqrt{3}}{2\pi}} \tag{3.8b}$$

where:
   as : replacement area ratio
   d : diameter of sand pile (m)
   D : interval of sand piles (m)
   V : volume of sand to be installed per unit depth (m^3)

   This procedure has been frequently adopted for its simplicity. However, as described above, it should be applied only to sandy grounds having fines content less than 20%. Figure 3.8 shows the relationship between fines content and SPT N-value between sand piles [6]. The figure clearly shows that the SPT N-value decreases with increasing fines content of the original ground. In the case of sandy grounds whose fines content exceeds 20%, procedure C or D should be adopted for design, which will be introduced later.

Procedure B
This procedure was proposed by Ogawa and Ishidou [8]. In this procedure, the void ratios of original and improved grounds are estimated by the SPT N-value where the effect of uniformity coefficient, U_c, and the overburden pressure are taken into account. The uniformity coefficient, U_c, is defined as D_{60}/D_{10}, where D_{60} and D_{10} are 60% and 10% diameters of soil particles respectively. However, the effects of upheaval and horizontal deformation of ground due to sand piles installation are not taken into

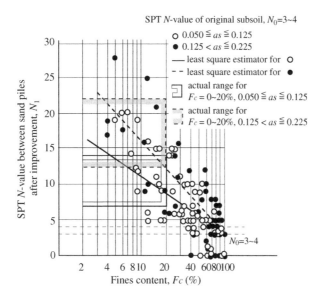

Figure 3.8.   Relationship between the SPT N-value of improved ground and fines content [7].

account in this procedure. Figures 3.9 and 3.10 show the relationships between the SPT N-value, relative density, and overburden pressure [8]. Figure 3.9 was presented by Ogawa and Ishidou [8] according to test data in the literatures [9], and they modified the D_{60} and e_{max} relation slightly to the broken lines as shown in the figure. Figure 3.10a was originally presented by Thornburn [10] based on the laboratory test data of Gibbs and Holtz [11].

In the procedure, the maximum and minimum void ratios, e_{max} and e_{min}, of the original ground are estimated at first. The magnitudes of e_{max} and e_{min} can be directly measured now by a standardized laboratory test [12]. However, there was no standardized testing method when this procedure was proposed. In this procedure, they are estimated by the relationship shown in Figure 3.9 together with the 60 percent diameter, D_{60}, and the uniformity coefficient, U_c, of the *in-situ* soil, which are measured in a laboratory particle distribution test.

The relative density of the original ground, e_i, corresponding to the SPT N-value is estimated by Figure 3.10a with the overburden pressure at the SPT measured depth. Then the void ratio of the original ground is calculated by Equation (3.9) together with the estimated e_{max} and e_{min} values and the relative density, D_r.

The void ratio corresponding to the target SPT N-value, e_t, is also estimated in a similar manner. In this phase, the overburden pressure to be considered is usually given as that at the measured depth or at the depth of 1/2 to 1/3 of the improved depth.

$$e = e_{max} - D_r \cdot (e_{max} - e_{min})$$

$$(3.9)$$

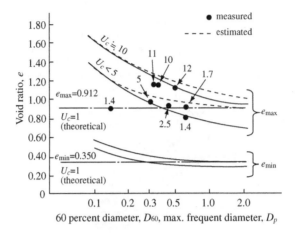

Figure 3.9.   Relationship between D_{60} and void ratio, e [8].

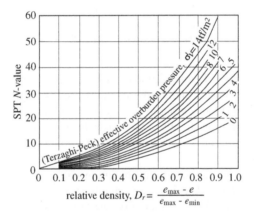

Figure 3.10a.   Relationship between relative density, D_r, and SPT N-value [8].

After obtaining e_i and e_t, the replacement area ratio, as, is calculated by Equation (3.10). The sand volume to be installed and the diameter of sand piles are obtained by substituting the calculated as value into Equations (3.7) and (3.8).

$$as = \frac{e_i - e_t}{1 + e_i} = \frac{\Delta e}{1 + e_i} \qquad (3.10)$$

where:

  as : replacement area ratio
  e_i : initial void ratio of original ground
  e_t : target void ratio of improved ground

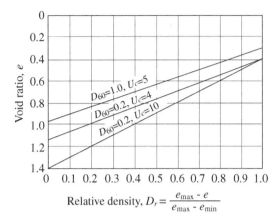

<figure>Relative density, $D_r = \dfrac{e_{max} - e}{e_{max} - e_{min}}$</figure>

Figure 3.10b.   Relationship between relative density, D_r, and void ratio, e [8].

Since the effect of fines content of the original ground is not directly taken into account, but is indirectly taken into account by D_{60} and U_c, care is required when applying this procedure to sandy grounds with relatively high fines content.

Procedure C
This procedure is the same as procedure B in principle except it incorporates the effect of fines content on improvement effect [7]. First, the maximum and minimum void ratios, e_{max} and e_{min}, of the original ground are estimated by the fines content, Fc, instead of Figure 3.9. Figures 3.11a and 3.11b show the laboratory test data on the relationship between e_{max} and e_{min} and Fc, which were presented by Hirama [13]. These data were measured by the laboratory test standardized by the Japanese Society of Soil Mechanics and Foundation Engineering [12]. Straight lines in the figure are a linear approximation of the data drawn by Mizuno et al. [7] for this design procedure, which are formulated as Equations (3.11a) and (3.11b) (The original formulation of e_{min} by Mizuno et al. [7] was mistaken and is corrected as Equation (3.11b)).

$$e_{max} = 0.02 \cdot Fc + 1.0 \qquad\qquad (3.11a)$$

$$e_{min} = 0.008 \cdot Fc + 0.6 \qquad\qquad (3.11b)$$

where:
  e_{min} : minimum void ratio
  e_{max} : maximum void ratio
  Fc : fines content (%)

The relative density of ground is estimated from the measured SPT N-value. There are various equations for relationship between the SPT N-value and relative density as explained later. In this procedure, the relationship proposed by Meyerhof [14] has

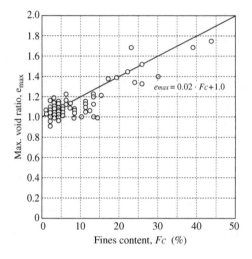

Figure 3.11a.   Relationship between maximum void ratio and fines content [7].

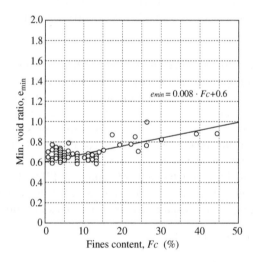

Figure 3.11b.   Relationship between minimum void ratio and fines content [7]. (The original formulation of e_{min} by Mizuno *et al.* was mistaken and is corrected as shown in the figure.)

been adopted as shown by Equation (3.12), which incorporates the effect of overburden pressure on the SPT N-value.

$$D_r = 21\sqrt{\frac{100 \cdot N}{70 + \sigma'_v}} \qquad (3.12)$$

where:

D_r : relative density (%)

N : SPT N-value of ground

σ_v' : overburden pressure at SPT measured depth (kN/m²)

Next, the void ratio of the original ground, e_i, is obtained by substituting the estimated e_{max}, e_{min} and D_r values into Equation (3.9).

The target void ratio of improved ground, e_t, can be obtained in the same manner according to the target SPT N-value, N_t. Again, the overburden pressure to be considered in Equation (3.12) is usually obtained as that at the measured depth or at the depth of 1/2 to 1/3 of the improved depth. The effect of fines content on the SPT N-value is taken into account in this procedure as follows. Accumulated field data have shown that the SPT N-value after SCP improvement is less than the predicted value in the case where the fines content of the original ground increases. Mizuno *et al.* [7] discussed the reason for the phenomenon as follows:

i) Equation (3.12) can underestimate the relative density calculated for sandy grounds with relatively high fines content. Therefore, the SPT N-value calculated by Equation (3.12) can be overestimated to the actual value.

ii) Sandy ground with high fines content can easily upheave and/or deform horizontally, which somewhat reduces the improvement effect.

A reduction factor, β, is introduced in this procedure to incorporate the effect of fines content on the SPT N-value, which is defined as Equation (3.13).

$$\beta = \frac{\Delta N}{\Delta N'} = \frac{N_t - N_i}{N_t' - N_i} \tag{3.13}$$

where:

N_i : SPT N-value of original ground

N_t : SPT N-value of improved ground incorporating fines content effect

N_t' : SPT N-value of improved ground without incorporating fines content effect

β : reduction factor

ΔN : actual increment of SPT N-value

$\Delta N'$: predicted increment of SPT N-value

According to accumulated field data, the relationship between the reduction factor, β, and fines content, Fc, is obtained as shown in Figure 3.12 [7]. The figure shows that the reduction factor, β, decreases linearly with increasing fines content, Fc, in a semi-logarithm scale, which is expressed as Equation (3.14) [7]. In the figure, a full line and two broken lines represent the linear approximation and the range of standard deviation respectively.

$$\beta = 1.05 - 0.51 \cdot \log_{10} Fc \tag{3.14}$$

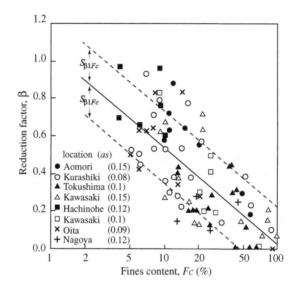

Figure 3.12.   Relationship between the fines content and reduction factor [7].

where:
  Fc : fines content (%)
  β : reduction factor

The target SPT N-value is obtained by considering the design SPT N-value and the effect of fines content. The target SPT N-value after improvement, N_t, can be obtained by Equation (3.15).

$$N_t = N_i + \frac{\left(N_t' - N_i\right)}{\beta} \qquad (3.15)$$

where:
  N_i : SPT N-value of original ground
  N_t : SPT N-value of improved ground incorporating fines content effect
  N_t' : SPT N-value of improved ground without incorporating fines content effect
  β : reduction factor

After obtaining the target SPT N-value by Equation (3.15), the relative density, D_r, and the target void ratio of improved ground, e_t, can be calculated in a similar manner to the original ground, which is calculated by Equations (3.12) and (3.9) together with the estimated e_{max} and e_{min}. Finally, the sand volume to be installed and the diameter of sand piles can be obtained by substituting the initial and target void ratios, e_i and e_t, into Equations (3.7) and (3.8).

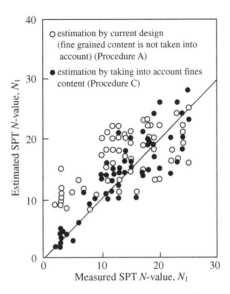

Figure 3.13.   Relationship between estimated and measured SPT N-values [7].

Figure 3.13 shows the relationship between estimated and field measured SPT N-values of improved ground [7]. The SPT N-values in the figure are normalized for an effective overburden pressure of $100 \, kN/m^2$ (1 kgf/cm²). In the figure, the estimations by Procedures A and C are plotted as open circles and full circles respectively. The figure clearly shows that the estimated SPT N-values by Procedure A overestimate the actual SPT N-values especially in grounds with relatively low SPT N-values. The estimations by Procedure C, on the other hand, coincide well with the field measured SPT N-value over a wide range of SPT N-values.

Procedure D
Procedures B and C assume that no ground upheaving (no volume change) takes place during sand piles installation. This assumption means that all sands installed in a ground function to increase ground density. However, there are many case histories where some amount of ground upheaving (volume increase) take place during SCP improvement and not all sands function to increase ground density. An example of field measurements on volumetric strain is shown in Figures 3.5a and 3.5b, which demonstrates this phenomenon. This phenomenon becomes dominant in grounds with high fines content.

Procedure D was proposed by Yamamoto *et al.* [15–16], in which a new parameter, Rc, called 'effective compaction ratio' is introduced to incorporate the ground upheaval effect. In the procedure, the maximum and minimum void ratios of the original ground, e_{max} and e_{min}, are estimated first by Equations (3.11a) and (3.11b) respectively, which is the same approach as Procedure C.

Table 3.1.   Increment of SPT N-value, ΔN_f [17].

Fines content, Fc (%)	ΔN_f
0 to 5	0
5 to 10	interpolate
higher than 10	0.1 Fc + 4

Table 3.2.   Increment of SPT N-value, ΔN_f [15].

Fines content, Fc (%)	ΔN_f
0 to 5	0
5 to 10	1.2 $(Fc - 5)$
10 to 20	6 + 0.2 $(Fc - 10)$
higher than 20	8 + 0.1 $(Fc - 20)$

Tokimatsu and Yoshimi [17] found from accumulated data that the SPT N-value of saturated sands with fines content exceeding 10% is 5 less than that of clean sands. They modified Equation (3.12) and proposed Equation (3.16) to estimate the relative density of ground by incorporating the fines content effect.

$$D_r = 21\sqrt{\frac{100 \cdot N}{70 + \sigma'_v} + \frac{\Delta N_f}{1.7}} \qquad (3.16)$$

where:

D_r : relative density (%)
N : SPT N-value of ground
ΔN_f : increment of SPT N-value for fines content effect
σ'_v : overburden pressure at SPT measured depth (kN/m^2)

They gave the magnitude of the increment of SPT N-value, ΔN_f, for several Fc values as shown in Table 3.1 [17]. Yamamoto *et al.* [15] slightly modified the above values and proposed ΔN_f as shown in Table 3.2.

Next, the void ratio of the original ground, e_i, is calculated by substituting the estimated relative density, D_r, and the maximum and minimum void ratios, e_{max} and e_{min}, into Equation (3.9).

Yamamoto *et al.* [15] analyzed the accumulated field data on liquefaction prevention and showed the relationship between effective compaction ratio, Rc, and fines content, Fc, as shown in Figures 3.14a and 3.14b. The field data of improved grounds by the vibrating compaction techniques are plotted in Figure 3.14a and those by the static compaction techniques (non-vibrating compaction techniques) are plotted in Figure 3.14b. They proposed the formulation for both techniques as Equation (3.17) [16].

$$Rc = 1.05 - 0.46 \cdot \log_{10} Fc \qquad (3.17)$$

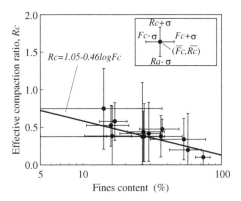

Figure 3.14a. Relationship between effective compaction ratio, Rc, and fines content, Fc for vibrating compaction techniques [15].

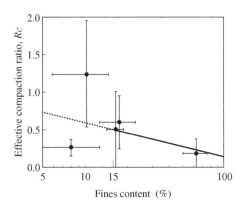

Figure 3.14b. Relationship between effective compaction ratio, Rc, and fines content, Fc for static compaction techniques [15].

where:
Fc : fines content (%)
Rc : effective compaction ratio

Beside this, the relative density, D_r, and void ratio of improved ground, e_t, are calculated by Equations (3.16) and (3.9) together with the target SPT N-value, N_t. Again, the magnitude of overburden pressure to be substituted in Equation (3.16) is usually assumed as that at the measured depth or at the depth of 1/2 to 1/3 of the improved depth. Then the replacement area ratio, as, is obtained by substituting the initial and target void ratios, e_i and e_t, and the effective compaction ratio, Rc, into Equation (3.18).

$$as = \frac{e_i - e_t}{Rc \cdot (1 + e_i)} \tag{3.18}$$

where:

    as : replacement area ratio
    e_i : initial void ratio of original ground
    e_t : target void ratio of improved ground
    Rc : effective compaction ratio

Finally, the sand volume to be installed and the diameter of compacted sand piles can be obtained by substituting the calculated replacement area ratio, *as*, into Equations (3.7) and (3.8).

3.2.4   Extent of SCP improved ground

Here, the design procedure for the extent of SCP improved ground for port structures is introduced as an example. The extent of SCP improved ground for liquefaction prevention should be designed according to the following considerations [1–3].

Propagation of excess pore water pressure into improved area
Excess pore water pressure is generated in an unimproved area and also in an improved area during an earthquake. The excess pore water pressure in the unimproved area is much higher than that in the improved area. The difference in pore water pressure drives propagation of the excess pore water pressure from the unimproved area into the improved area. Shaking table tests and seepage flow analyses suggest that the excess pore water pressure within the area defined by the rectangle ABCD in Figure 3.15 increases to higher than 0.5 σ'_v in the case when the unimproved area is liquefied, where σ'_v is an effective vertical stress. As a result, the shear strength of the soil in the area ABCD in Figure 3.15 reduces considerably. This indicates that the extent of improved area should be wide enough to compensate this area.

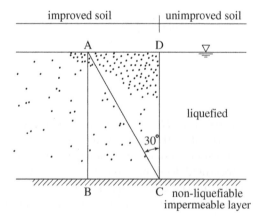

Figure 3.15.   Area influenced by propagation of pore water pressure.

Water pressure by liquefied area

During an earthquake excitation, both static and dynamic water pressures act on the boundary of the liquefied and non-liquefied area as shown in Figures 3.16a and 3.16b. Figures 3.16a and 3.16b show the cases when the earthquake excitation acts toward the improved area or the unimproved area respectively. The magnitudes of static and dynamic water pressures are calculated by Equations (3.19a) and (3.19b) respectively.

static water pressure:

$$p_{ws} = \gamma' \cdot H \qquad\qquad (3.19a)$$

dynamic water pressure:

$$p_{wd} = \frac{7}{8} \cdot \frac{a}{g} \cdot \gamma_{sat} \sqrt{\gamma \cdot z} \qquad\qquad (3.19b)$$

where:

a : acceleration (gal)
g : acceleration due to gravity (gal)
p_{wd} : dynamic water pressure (kN/m²)
p_{ws} : static water pressure (kN/m²)
z : depth (m)
γ' : effective unit weight of ground (kN/m³)
γ_{sat} : saturated unit weight of ground (kN/m³)

The extent of improved ground should be wide enough to prevent failure or significant deformation of the structure by these water pressures. Figure 3.17 shows a schematic diagram to investigate the stability of the retaining structure [3]. In the stability analysis,

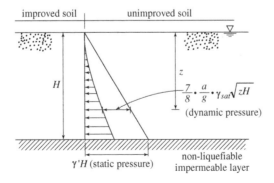

Figure 3.16a.   Water pressure acting on boundary of improved ground at earthquake excitation toward unimproved area [3].

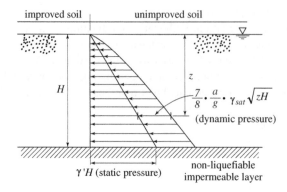

Figure 3.16b.   Water pressure acting on boundary of improved ground at earthquake excitation toward improved area [3].

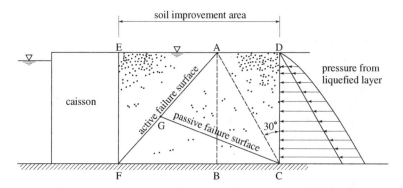

Figure 3.17.   Schematic diagram for stability calculation of retaining structure [3].

active and passive failure surfaces are assumed and the propagation of water pressure is also incorporated.

Shear strength reduction of liquefied area
When soils are liquefied, the shear strength reduces considerably and is usually assumed to be zero in practical designs. The shear strength in the area ABCD in Figure 3.17 is also assumed to be zero due to the pore water pressure propagation as mentioned in (1). The improved area should be wide enough to assure bearing capacity by the shear strength mobilized in the area EFBA in Figure 3.17. In the figure, the earth pressure acting on the boundary is reduced by subtracting the dynamic water pressure from the static earth pressure for a safe calculation.

According to the above-mentioned considerations, the extents of improved ground are exemplified as shown in Figure 3.18 for several types of structures [3].

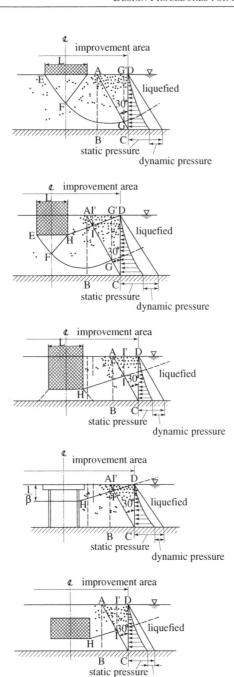

Figure 3.18.   Illustration for extent of SCP improvement [3].

REFERENCES

1   Ministry of Transport: Technical standards and commentaries for port and harbour facilities in Japan. Ministry of Transport, Japan, 1999 (in Japanese).
2   The Overseas Coastal Area Development Institute of Japan: English version of technical standards and commentaries for port and harbour Facilities in Japan. 2002.
3   International Navigation Association: Seismic design guidelines for port structures. 2001.
4   Yamamoto, M., Harada, K., Nozu, M. and Ohbayashi, J.: Study on evaluation of increasing density due to penetrating sand piles, Proc. of the 32nd Annual Conference of the Japanese Society of Soil Mechanics and Foundation Engineering, pp.2631–2632, 1997 (in Japanese).
5   Japanese Society of Soil Mechanics and Foundation Engineering: Soil improvement methods – survey, design and execution–. The Japanese Society of Soil Mechanics and Foundation Engineering, 1988 (in Japanese).
6   Fudo Construction Co., Ltd.: Design manual for compozer system. 1971 (in Japanese).
7   Mizuno, Y., Suematsu, N. and Okuyama, K.: Design method for sand compaction pile for sandy soils containing fines. Journal of the Japanese Society of Soil Mechanics and Foundation Engineering, 'Tsuchi-to-Kiso', Vol.35, No.5, pp.21–26, 1987 (in Japanese).
8   Ogawa, M. and Ishidou, M.: Application of compozer method for sandy ground. Journal of the Japanese Society of Soil Mechanics and Foundation Engineering, 'Tsuchi-to-Kiso', Vol.13, No.2, pp.77–81, 1965 (in Japanese).
9   Hutchinson, B. and Townsend, D.: Some grading-density relationships for sands. Proc. of the 5th Internal Conference on Soil Mechanics and Foundation Engineering, Vol.1, pp.159–163, 1961.
10 Thornburn, S.: Tentative correction chart for the standard penetration test in non cohesive soils. Civil Engineering & Public Works Review, Vol.58, No.683, pp.752–753, 1963.
11 Gibbs, H.J. and Holtz, W.G.: Research on determining the density of sands by spoon penetration testing. Proc. of the 4th Internal Conference on Soil Mechanics and Foundation Engineering, Vol.1, pp.35–39, 1957.
12 The Japanese Society of Soil Mechanics and Foundation Engineering: The method for minimum and maximum densities of sands, JSG 0161. The Japanese Society of Soil Mechanics and Foundation Engineering, 2000 (in Japanese).
13 Hirama, K.: On application of relative density of sand. Proc. of the Symposium on Relative Density and Mechanical Property of Sand, pp.53–56, 1981 (in Japanese).
14 Meyerhof, G.G.: Discussion of Session 1. Proc. of the 4th Internal Conference on Soil Mechanics and Foundation Engineering, Vol.3, p.110, 1957.
15 Yamamoto, M., Sakai, S., Nakasumi, I., Higashi, S., Nozu, M. and Suzuki, A.: Estimation of improvement effects on sandy ground by sand compaction pile.

Proc. of the 32nd Annual Conference of the Japanese Society of Soil Mechanics and Foundation Engineering, pp.2315–2316, 1997 (in Japanese).

16 Yamamoto, M., Harada, K. and Nozu, M.: New design of sand compaction pile for preventing liquefaction in loose sandy ground. Journal of the Japanese Society of Soil Mechanics and Foundation Engineering, 'Tsuchi-to-Kiso', Vol.48, No.11, pp.17–20, 2000 (in Japanese).

17 Tokimatsu, K. and Yoshimi, Y.: Empirical correlation of soil liquefaction based on SPT N-value and fines content. SOILS AND FOUNDATIONS, Vol.23, No.4, pp.56–74, 1983.

Chapter 4

Execution, Quality Control and Assurance

4.1 INTRODUCTION

Several types of SCP machine have been developed and adopted for on-land construction and marine construction for more than 50 years, which include hammering
compaction technique, vibrating compaction technique and static compaction technique (non-vibrating compaction technique). In this section, the SCP machine and
execution of the vibrating compaction technique, which has been used in many construction sites, are briefly introduced as well as the quality control and assurance in
SCP improvement. The SCP machine and execution of the other techniques are
described in Chapter 6 in detail as well as the historical background of machinery
development.

4.2 MATERIALS

As described in Chapter 2, the sand piles for the SCP method are expected to function
as a stiff material for suspending an external load and as a drainage material, which
requires a well blended granular material. The granular material suitable for the SCP
method is soils with low fines content and with high particle strength so that negligible crushing of soil particles takes place during the execution. Figure 4.1 shows the
accumulated case histories of soil particle distributions applied to the SCP method
[1–2]. The figure indicates that soils with fines content of 3% to 5% and maximum
particle size of 40 mm to 50 mm have been used for the method. According to the
accumulated experiences, soils whose particle size distribution falls within the ranges
in the figure have sufficient shear strength and permeability for the method. In the
case where suitable soil cannot be obtained due to economic and/or environmental
reasons, a soil with poor quality can be used after confirming acceptable performance
and execution ability. There are some case histories where soils with fines content of
10% to 15% were used in the SCP method with a high replacement area ratio without
expectation of drainage function.

Recently, the limited availability of sands suitable for the method has necessitated
research to find new materials. As a result of such research, many new materials including steel, copper and ferro-nickel slag, oyster shell and granulated coal fly ash have

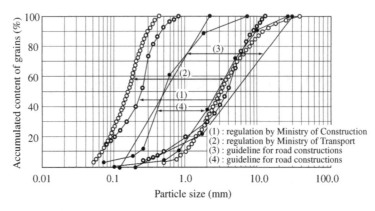

Figure 4.1.   Soil particle distribution suitable for SCP method [1–2].

been used as SCP materials as described later in Chapter 6. Two case histories where copper slag or oyster shell was used as a SCP material are introduced in Chapter 5.

4.3   SCP MACHINES

Several types of SCP machine have been developed and adopted for on-land construction and marine construction for more than 50 years, which will be introduced in Chapter 6 in detail. However, some of them were replaced by new techniques, such as the hammering compaction technique that was the first development of the SCP method. Techniques for executing the SCP method can be divided into three groups as shown later in Table 6.2: hammering compaction technique, vibrating compaction technique and static compaction technique (non-vibrating compaction technique). In the hammering compaction technique, sand fed out from a casing pipe is compacted in a similar manner to the pile driving technique, by dropping a heavy weight (ram) onto the sand in the casing pipe. In the vibrating compaction technique, sand fed into a ground is compacted by vibratory excitation. The vibrating compaction technique has three major variations according to the position and type of vibration. This technique has been frequently applied for both on-land construction and marine construction. In the static compaction technique (non-vibrating compaction technique), on the other hand, sand introduced in a ground is compacted by static rotational and/or downward movements of a casing pipe or a probe. This technique has three variations according to the type of compaction. Since this technique has several advantages including quite low noise and vibration, it has been applied to on-land construction especially in urban areas, and recently to marine construction [3].

In this section, the vibrating compaction technique, which has been used in many construction sites, is introduced. The other techniques will be introduced in Chapter 6 as well as the historical background of machinery development.

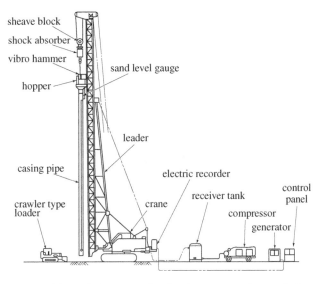

Figure 4.2a.   SCP machine of the vertical vibrating compaction technique for on-land construction.

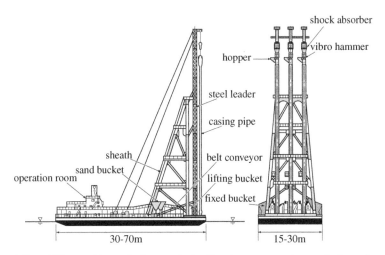

Figure 4.2b.   SCP machine of the vertical vibrating compaction technique for marine construction.

The vertical vibrating compaction technique in the vibrating compaction technique has been most frequently used in SCP improvement. Typical SCP machines of the vertical vibrating compaction technique are shown in Figures 4.2a and 4.2b for on-land construction and marine construction respectively.

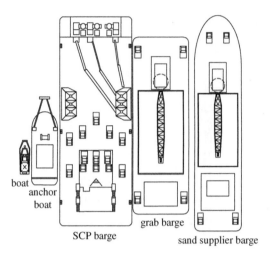

boat

anchor
boat

grab barge

SCP barge

sand supplier barge

Figure 4.3.   SCP barges for marine construction.

The machine for on-land construction consists of a crawler crane with a leader and a casing pipe, a vibro-hammer and a compressor to supply high air pressure to a casing pipe. A crawler crane with a lifting capacity of 250 kN to 400 kN (25 tf to 40 tf) is often used. The casing pipe of 40 cm to 50 cm in diameter is suspended along the leader through the vibro-hammer to construct compacted sand piles whose diameter ranges from 50 cm to 70 cm. The vibro-hammer is suspended from the crane via a shock absorber. A lifting bucket is also suspended along the leader and goes up and down to supply granular material to the casing pipe through the hopper.

In marine construction, a special barge equipped with SCP machine is used to construct compacted sand piles *in situ* (Figure 4.2b). The SCP machines for marine construction usually have more than three leaders and casing pipes. A casing pipe with a diameter of 80 cm to 120 cm is suspended along a leader through a vibro-hammer. Figure 4.3 shows a group of barges for marine construction. The granular material is transported from the sand supplier barge to the SCP barge through the grab barge. The grab barge functions as a kind of buffer for sand material. The material transported to the SCP barge is transported by means of belt conveyors and then is fed to the casing pipes by the lifting bucket. The SCP machines currently available for marine construction are capable of constructing sand piles whose diameter ranges from 100 cm to 200 cm and the maximum depth of pile reaches up to 70 m from sea level.

4.4   EXECUTION PROCEDURE

A typical construction procedure is shown in Figure 4.4, which includes preliminary survey of ground, field trial test, positioning, operation and monitoring.

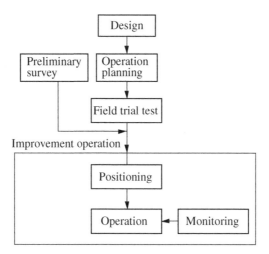

Figure 4.4.   Typical SCP construction procedure.

Preliminary survey
Before actually executing ground improvement, execution circumstances should be checked to ensure sufficient quality of improved ground, smooth operation and to prevent adverse environmental impact. The execution schedule can be disturbed by severe weather and wave conditions in marine construction, so weather and wave conditions should be examined in advance when making the execution schedule; wave height, wind direction, wind velocity and tides should be carefully surveyed. According to previous case records, marine construction is difficult to conduct in conditions where the maximum wind velocity exceeds 10 m/second, the maximum significant wave height exceeds 80 cm, or the minimum visibility is less than 1,000 m. Adverse environmental impacts such as noise and vibration, etc., which can occur during the execution should obviously be kept to a minimum.

In the SCP method, a casing pipe with large diameter is penetrated into a soft ground. Any obstacles on or below the ground in the construction area can delay the operation schedule, or cause damage to the casing pipe. Before execution, the ground should be surveyed carefully and any obstacles should be removed. This process is particularly important for marine construction with regard to blind shells that can cause human injury. The soil survey in marine construction can usually be carried out by means of a magnetic prospecting probe.

Field trial test
It is advisable to conduct a field trial test in advance at a ground or adjacent to the construction site, in order to ensure the quality of improvement and smooth execution at the site. In the test, all the monitoring equipment, such as the level gauges for casing pipe and sand surface, and the depth gauge and load gauge should be calibrated.

Improvement operation (positioning processing)
The procedure for constructing compacted sand piles by the vertical vibrating compaction technique is shown in Figure 4.5. In the stage of setting the SCP machine, the casing pipe should be correctly positioned according to the design. There are four techniques for positioning the SCP machine for marine works: collimation by two transit apparatuses, collimation by a transit and an optical range finder, an automatic positioning system with three optical finders, and a positioning system with GPS. After setting the machine, a casing pipe is penetrated to a prescribed depth. The casing pipe is then retrieved to ground level, while sand or granular material is fed and compacted in the ground to construct stiff sand piles.

i)   The casing pipe is located at the design position.
ii) The casing pipe is driven into the ground with the help of vertical vibratory excitation by the vibro-hammer on the top of the casing pipe. Compressed air is injected from the outlet nozzles installed on the side face of the casing pipe in the case of penetrating relatively hard stratum. During the penetration, the casing pipe is filled with SCP material that is supplied through the hopper at the upper end of the casing pipe by the lifting bucket.
iii) After reaching the prescribed depth, the casing pipe is retrieved about one meter to feed the sand in the casing pipe into the ground with the help of compressed air.
iv) The sand fed into the ground is compacted to expand the diameter by the vibratory excitation of the casing pipe in the vertical direction. The degree of compaction is controlled so that the diameter of the sand pile becomes the designed value.

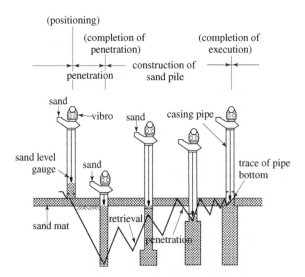

Figure 4.5.   Execution procedure for vertical vibrating compaction technique.

v) After compacting the sand to the designed degree, the casing pipe is retrieved about one meter again and the sand is fed into the ground. The sand is compacted again by the vibratory excitation.

vi) These procedures are repeated until a compacted sand pile is constructed up to ground level. During the procedure, SCP material is continuously supplied through the hopper by the lifting bucket. During the construction, the depth of the casing pipe and the position of the sand in the casing pipe are continuously measured for quality control and assurance.

Operation monitoring (quality control/quantity control)
To assure the quality and dimensions of compacted sand piles, it is essential to maintain the designed condition during the actual execution by monitoring the quantity of sand material, the levels of casing pipe and sand surface, etc. These monitoring data can be fed back to the operator for precise construction of sand piles. The depth gauge (GL) to indicate the position of the casing pipe, and the sand level gauge (SL) to indicate the sand level in the casing pipe are absolutely essential. The GL gauge consists of a drum that rotates as the casing pipe moves. Its rotational displacement is converted into the position of the casing pipe. The SL gauge measures the sand level in the casing pipe by a weight electrode. These data are continuously measured and recorded to assure precise operations.

4.5   QUALITY CONTROL

4.5.1   Quality control monitoring items

Quality control monitoring items are summarized in Table 4.1 [4], which are categorized into before, during and after executions.

Before the execution, the particle size distribution test should be carried out to measure the property of the sand material and to assure that the material is suitable for the SCP method. For SCP applications to sandy grounds, the fines content of the original ground should be measured, as it influences the improvement effect as described in Chapter 3. Also, the machinery should be checked and all the monitoring gauges should be calibrated. In marine construction, the sea depth should be measured by sounding techniques.

During the execution, the location of each sand pile should be precisely positioned and recorded. It is essential to maintain the designed condition during the actual execution by monitoring and controlling the length of sand piles, the volume of sand fed into the ground, and the continuity of sand piles to assure the quality and dimensions of compacted sand piles. Beside the monitoring items listed in Table 4.1 [4], the verticality of sand piles and the soil disturbance of cohesive soil are measured in some cases. In marine construction, the tidal level and the sea depth should be monitored.

After the execution, a series of standard penetration tests is usually conducted to confirm the strength and continuity of sand piles. In the case where the shear strength

Table 4.1.   Monitored and controlled items for quality control [4].

Period	Quality control items	Control items
Before execution	• property of sand material • confirmation of machinery • calibration of gauges • level of sea depth	• measurement of particle size distribution • sounding of sea bottom
During execution	• length of sand pile, volume and continuity of sand pile • location of sand pile • level of sand pile • property of material	• quality control during execution • collimation by two transit apparatuses, collimation by a transit and an optical range finder, an automatic positioning system with three optical finders, and a positioning system with GPS • tide level, sea depth • measurement of particle size distribution
After execution	• strength and continuity of sand pile • depth of sea bottom • upheaval volume	• standard penetration test • sounding of sea bottom

decreases and the recovery phenomenon are taken into account in the design, the shear strength of clay ground as well as the ground level of the upheaval portion should be measured.

Figure 4.6 summarizes the items to be monitored and controlled during the execution. Regarding the quality of sand piles, the volume of sand and the compacting are checked by monitoring and controlling the depth gauge and the sand level gauge and level gauge. The location and length of sand piles are checked by transit or GPS, and by depth gauge and load gauge.

4.5.2   Control of sectional area of sand pile in ground

One of the most important quality control items during the execution is supplying the designed volume of granular material into the ground and compacting it to the designed degree. The quality control regarding these is explained in detail as follows. Figure 4.7 shows an illustration of a sand pile during execution by the vertical vibrating compaction technique. However, this explanation is applicable also to other execution techniques.

The weight of sand in a casing pipe is estimated by the measured levels of casing pipe and sand, CLG_0 and SLG_0, by Equation (4.1). The volume of sand in the casing pipe is usually lower (density of sand in the casing pipe is higher) than that before filling due to the vibration effect of the casing pipe during penetration. The volume change ratio of sand in the casing pipe before and after filling is introduced as a

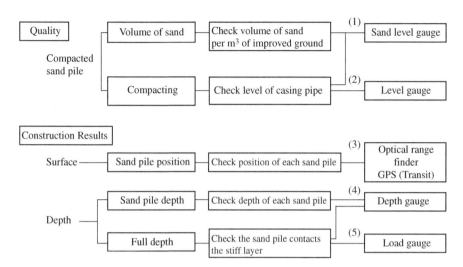

Figure 4.6.   Items of improvement monitoring.

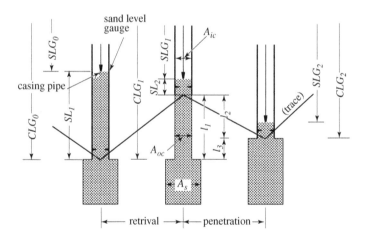

Figure 4.7.   Quality control in manufacturing compacted sand pile.

parameter, Rv_1, which is defined as Equation (4.2). The volume change ratio is inverse relation to the density ratio before and after filling the casing pipe. The magnitude of the volume change ratio, Rv_1, is usually considered to be around 1.3 according to previous case histories.

When the casing pipe is retrieved to the level of CLG_1, the weight of sand fed into the ground is obtained by Equation (4.3). At this time, the level of the sand remaining in the casing pipe is measured as SLG_1. It is usual that the sand level in the casing pipe before retrieval, SLG_0, is not equal to SLG_1, the sum of the levels of SL_2 and l_1, because of the difference of the inner and outer diameters of the casing pipe. As the sand in the casing pipe is in a dense condition, it is assumed that the density of sand does not change during feeding into the ground. When the casing pipe penetrates again to the depth of CLG_2, the sand fed into the ground is compacted and its sectional area becomes A_s. By introducing a parameter, R_{v2}, as a volume change ratio due to compaction, which is defined as Equation (4.4), the sectional area of the sand pile after compaction, A_s, is easily calculated by Equation (4.5). The ratio, R_{v2}, has been obtained as 1.0 to 1.3 according to previous case histories.

$$\begin{aligned} W_{sand} &= \gamma_{sand\ in\ casing\ pipe} \cdot A_{ic} \cdot (CLG_0 - SLG_0) \\ &= \gamma_{sand\ in\ casing\ pipe} \cdot A_{ic} \cdot SL_1 \end{aligned} \tag{4.1}$$

$$\begin{aligned} R_{v1} &= \frac{V_{sand\ before\ filling}}{V_{sand\ after\ filling}} \\ &= \frac{\gamma_{sand\ after\ filling}}{\gamma_{sand\ before\ filling}} \end{aligned} \tag{4.2}$$

$$\begin{aligned} W_{sand} &= \gamma_{sand\ in\ casing\ pipe} \cdot A_{ic} \cdot \{(CLG_0 - SLG_0) - (CLG_1 - SLG_1)\} \\ &= \gamma_{sand\ in\ casing\ pipe} \cdot A_{ic} \cdot \{SL_1 - SL_2\} \\ &= \gamma_{sand\ in\ ground} \cdot A_{oc} \cdot l_1 \end{aligned} \tag{4.3}$$

$$\begin{aligned} R_{v2} &= \frac{V_{sand\ after\ feeding}}{V_{sand\ after\ compaction}} \\ &= \frac{\gamma_{sand\ after\ compaction}}{\gamma_{sand\ after\ feeding}} \end{aligned} \tag{4.4}$$

$$A_s = \frac{1}{Rv_2} \cdot \frac{A_{ic} \cdot \{(CLG_0 - SLG_0) - (CLG_1 - SLG_1)\}}{(CLG_0 - CLG_2)} \tag{4.5}$$

where:

A_{ic} : inner sectional area of casing pipe
A_{oc} : outer sectional area of casing pipe
A_s : sectional area of sand pile after compaction
CLG_0 : level of casing pipe before retrieval (m)
CLG_1 : level of casing pipe after retrieval (m)
CLG_2 : level of casing pipe after re-penetration (m)
R_{v1} : density ratio in casing pipe after and before vibration (usually around 1.3)
R_{v2} : density ratio of sand before and after in-situ compaction (usually 1.0 to 1.3)
SL_1 : length of sand in casing pipe before retrieval (m)
SL_2 : length of sand in casing pipe after retrieval (m)

SLG_0 : level of sand in casing pipe before retrieval (m)
SLG_1 : level of sand in casing pipe after retrieval (m)
SLG_2 : level of sand in casing pipe after re-penetration (m)

While manufacturing sand piles *in situ*, the level of the casing pipe, *CLG*, and the sand level in the casing pipe, *SLG*, are continuously monitored and displayed to the operator on the SCP machine. Figure 4.8 shows a typical example of an execution record. In the figure, the sand level in the casing pipe is varied from 0 to 6.6 m, which means that the sand is supplied into the pipe in several steps while constructing the sand pile. It can be seen that the casing pipe is penetrated almost monotonically to the prescribed depth of about 32 m. The casing pipe is then retrieved and penetrated several times to the depth of 20 m, where a compacted sand pile is constructed in the ground. The total time taken to construct the sand pile is about 28 minutes in this example.

4.5.3   Bottom treatment

To increase the stability of improved ground and/or to reduce the ground settlement, sand piles should preferably reach a stiff layer (fix type improvement). It is common that the depth of the stiff layer has local undulations and is sometimes different from the design value estimated by previous soil surveys. Therefore it is essential in an actual execution to confirm the depth of the stiff layer for every sand pile and to adjust the pile length based on the measured depth of the stiff layer. In the execution, the rapid change in penetration velocity of the casing pipe is one of the key indicators to determine whether the casing pipe has reached the stiff layer or not (see Figure 4.8).

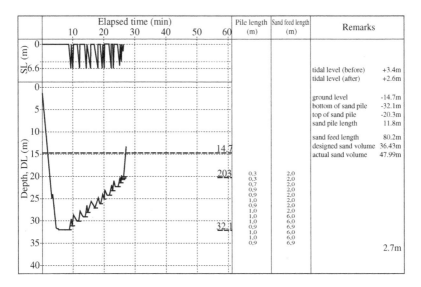

Figure 4.8.   A typical example of an execution record.

It is advisable to perform a field trial to calibrate the relationship between penetration speed of casing pipe and stiffness of ground.

4.6   QUALITY ASSURANCE

4.6.1   Frequency of standard penetration tests

After execution, a series of standard penetration tests (SPT) is often conducted to confirm the improvement effect. In SCP applications to clay ground, the tests are usually conducted at the center of sand piles to measure the strength and continuity of compacted sand piles. In SCP applications to sandy ground, on the other hand, the tests are often conducted at the point furthest from the sand piles (at the ground between sand piles) where the improvement effect is estimated to be minimum. The frequency of SPT tests should be determined according to the size of construction, number of SCP machines operated, complexity of ground condition and importance of the superstructure. Figure 4.9 shows the statistical records regarding the relationship between the total number of sand piles and frequency of SPT tests [4]. The frequency on the vertical axis is plotted as the ratio of the number of sand piles per one SPT test. There is much scatter in the accumulated data depending upon each structure and ground conditions, but the ratio of SPT tests increases with increasing total number of sand piles, which means lower frequency of SPT tests with increasing construction size. In general, SPT tests are carried out every 200 to 500 sand piles irrespective of the kind of application.

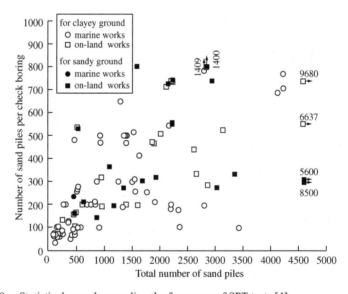

Figure 4.9.   Statistical records regarding the frequency of SPT tests [4].

4.6.2   Internal friction angle of compacted sand piles

The internal friction angle of compacted sand piles, ϕ, is usually estimated by the measured SPT N-value. There are many proposed relationships as summarized in Figure 8.15. Among them, the relationship proposed by Dunham as shown in Equation (4.6) has usually been adopted in design procedures [1–2]. The SPT N-value of at least 15 is usually required for application of clay ground to satisfy the designed internal friction angle of sand piles exceeding 30 or 35 degrees.

$$\phi = \sqrt{12 \cdot N} + 15 \tag{4.6}$$

where:

N : SPT N-value

ϕ : internal friction angle (degree)

4.7   NOISE AND VIBRATION DURING EXECUTION

The noise and vibration caused by SCP execution with distance from the machine are shown in Figures 4.10a and 4.10b respectively [4]. The figures show data of some soil improvement techniques including the Deep Mixing method [5] and the Vibro-flotation method. Among these improvement techniques, the noise and vibration levels

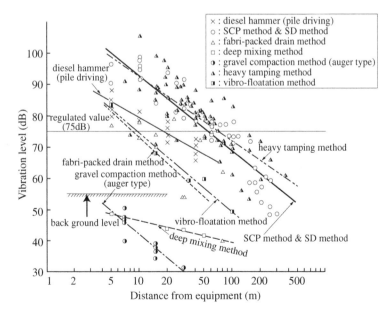

Figure 4.10a.   Vibration measurements during execution [4].

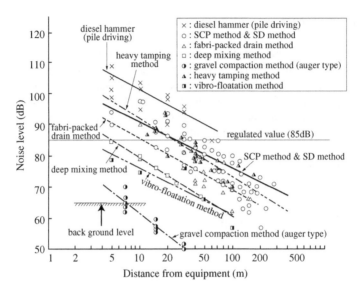

Figure 4.10b.   Noise measurements during execution [4].

caused by the SCP method are relatively high and decrease almost linearly with increasing logarithm of the distance. This means that sufficient care should be paid to the noise and vibration during execution, especially in urban areas. It is preferable to use the static compaction technique in urban areas, as it can reduce the noise and vibration considerably (see Chapter 6).

REFERENCES

1   Ministry of Transport: Technical standards and commentaries for port and harbour facilities in Japan. Ministry of Transport, Japan, 1999 (in Japanese).
2   The Overseas Coastal Area Development Institute of Japan: English version of technical standards and commentaries for port and harbour facilities in Japan. 2002.
3   Otsuka, M. and Isoya, S.: SAVE marine method. Journal of Construction Machine, No.8, pp.58–61, 2003 (in Japanese).
4   Japanese Society of Soil Mechanics and Foundation Engineering: Soil improvement methods – survey, design and execution–. The Japanese Society of Soil Mechanics and Foundation Engineering, 1988 (in Japanese).
5   Coastal Development Institute of Technology: The deep mixing method – principle, design and construction–. A.A. Balkema Publishers, 123 p., 2002.

Chapter 5

Case Histories

5.1 INTRODUCTION

From many previous SCP applications in Japan, some case histories are selected and briefly introduced in this Chapter. The case histories are summarized in Table 5.1 and Figure 5.1, and include several kinds of original ground condition, purpose of improvement, SCP material and machine.

Table 5.1. Case histories of SCP applications introduced.

Application	Type of ground	Construction site	Purpose of improvement	Location
Stability of embankment and reduction of settlement	peat	on-land	stability of embankment and reduction of settlement	Ebetsu
Bearing capacity of improved ground with low replacement area ratio	clay	marine	field loading test on bearing capacity	Maizuru Port
Horizontal resistance of pile structure	clay	marine	horizontal resistance of pile structure	Trans-Tokyo Bay Highway
Stability of steel cell type quay	clay	marine	stability of steel cell type quay	Kansai International Airport
Copper slag as SCP material	clay	marine	bearing capacity	Uno Port
Oyster shell as a SCP material	clay	marine	bearing capacity	Ishinomaki Port
Liquefaction prevention	sand	on-land	liquefaction prevention	Kushiro Port
Static compaction technique	sand	on-land	liquefaction prevention	Matsusaka Port

Figure 5.1.   Location of case histories introduced.

5.2   STABILITY OF EMBANKMENT AND REDUCTION OF SETTLEMENT

The Hokkaido Highway Construction Project was planned to improve the traffic connection between local towns in Hokkaido, for a part of which a highway embankment was constructed from Sapporo to Asahikawa (Figure 5.2). Thick soft peat ground was stratified there. As the peat ground was very soft with a high water content, it was anticipated to cause difficult problems of settlement and stability of the highway embankment. To address these, soil improvement was required. In Ebetsu, a local town midway between Sapporo and Asahikawa, field loading tests were carried out in 1978 to 1980 to investigate the applicability of various soil improvement methods to the peat ground, where full-scale embankments were constructed on sand drain and sand compaction pile improved grounds as well as an unimproved ground [1–2].

Figure 5.3 shows a soil profile at the test site [1–2]. It can be seen in the figure that a peat layer was stratified at the ground surface about 6 m in thickness, and had quite a high water content ranging from 400% to 1,000% and low strength of around 10 kN/m^2 to 20 kN/m^2 in unconfined compressive strength. Beneath the peat layer, soft clay layers were stratified alternately with depth, whose shear strength increased gradually with depth.

In the field test, three test embankments were constructed on the three types of grounds as shown in Figures 5.4a to 5.4c, which included no improvement area, a sand drain improved area and a sand compaction pile improved area [2]. In the ground

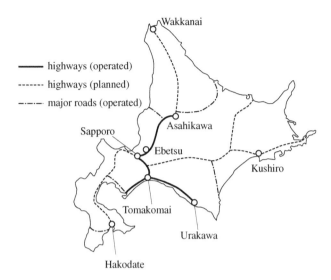

Figure 5.2.   The Hokkaido Highway Construction Project.

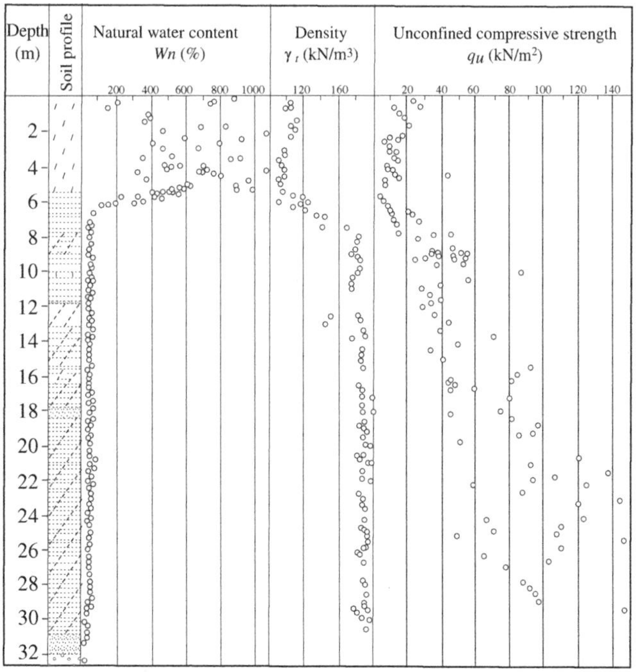

Figure 5.3.   Soil profile at test site [1–2].

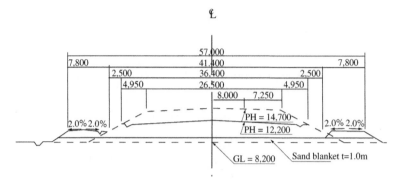

Figure 5.4a.   Schematic view of test embankment in no treatment area [2].

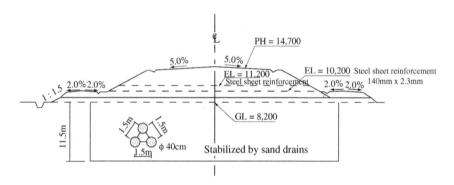

Figure 5.4b.   Schematic view of test embankment in sand drain improved area [2].

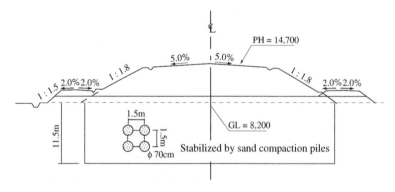

Figure 5.4c.   Schematic view of test embankment in sand compaction pile improved area [2].

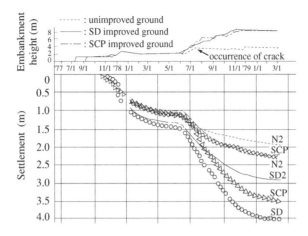

Figure 5.5.   Embankment height and ground settlement with elapsed time [2].

improved by the Sand Drain method, sand piles with a diameter of 40 cm were installed at a drain arrangement of equilateral triangular pattern with an interval of 1.5 m. The sand piles were constructed up to a depth of 11.5 m, which did not reach the stiff layer. After the soil improvement, the embankment was constructed on the ground in a step-wise manner to a height of 8 m. Two layers of steel mesh with 140 mm in width and 2.3 mm in thickness were extended within the embankment to reinforce the embankment. In the SCP improved area, the ground was improved with compacted sand piles with a diameter of 70 cm, and with a sand pile arrangement of square pattern with an interval of 1.5 m. The compacted sand piles were constructed up to a depth of 11.5 m, which did not reach the stiff layer. After the soil improvement, the embankment was constructed in a step-wise manner to a height of 8 m. In this area, no steel mesh was extended within the embankment.

Figure 5.5 shows the embankment height and ground settlement with elapsed time [2]. The ground settlement took place on the three types of ground as the embankments were constructed. For the unimproved ground (N2 in the figure), the ground settlement took place gradually during construction of the embankment and the consolidation period, and the amount of settlement was smaller than those of the improved grounds probably because negligible consolidation proceeded during the construction of embankment. The embankment failed with a crack and large horizontal deformation when the embankment height reached 3.5 m in mid July. After the failure, ground settlement continued to occur even though no further filling was carried out. For the grounds improved by the SD and SCP methods, on the other hand, the embankment was successfully constructed to the final height of 8 m. Upon completion of embankment construction, a relatively large settlement of more than 2 m was observed at these grounds. The final ground settlement was estimated by fitting a hyperbolic curve to the measured data, as summarized in Table 5.2 [2]. The estimated

Table 5.2.   Estimated final ground settlement [2].

Area	Embankment height	Estimated final settlement by hyperbola method
NT	3.5 m	234 cm
SD+RF	8.0 m	299 cm
SCP	8.0 m	224 cm

ground settlement of the improved grounds was 299 cm for the SD ground and 224 cm for the SCP ground. The ground settlement of the SCP improved ground was about 75% of that of the SD improved ground, which was due to the stress concentration on compacted sand piles in the SCP method. By assuming that the amount of ground settlement of the SD improved ground was equal to that of the unimproved ground, the settlement reduction factor, β, of 0.75 was obtained as the ratio of the settlement of SCP ground to that of unimproved ground.

The stress concentration was investigated at the SCP improved ground by the earth pressure measurements. The measured earth pressures on the sand pile head and on the surrounding soil showed that the embankment pressure concentrated on the sand piles as the construction proceeded. The stress concentration ratio, which was defined as the ratio of vertical stress acting on the sand pile to that acting on the surrounding soil, was obtained as about 3 by the measured earth pressures, which was quite close to the value back calculated by the settlement reduction factor, β.

5.3   BEARING CAPACITY OF IMPROVED GROUND WITH A LOW REPLACEMENT AREA RATIO

The SCP method with a relatively high replacement area ratio has been adopted for constructing port facilities such as sea revetments and breakwaters in which almost all soft soil is replaced by compacted sand piles. However, the method with a high replacement area ratio is quite expensive and causes the problem of how to dispose of the upheaval soil. Upheaval soil is usually excavated and dumped at disposal sites, which incurs extra expense. In order to make the method more economical, the SCP method with a low replacement area ratio is preferable. In order to investigate the applicability of the method, an *in-situ* full-scale loading test on SCP improved ground with a relatively low replacement area ratio was conducted at Maizuru Port, Kyoto Prefecture by the Third Bureau of Port Construction, Ministry of Transport in 1986 to 1988. In the test, many characteristics of the improved ground were investigated before, during and after the loading test, including the shape of upheaval ground, reduction and recovery in shear strength due to sand piles installation, stress concentration, and bearing capacity of improved ground. As many researches were conducted on the test data and presented elsewhere (e.g. [3–7]), only a brief outline of the test results is given here.

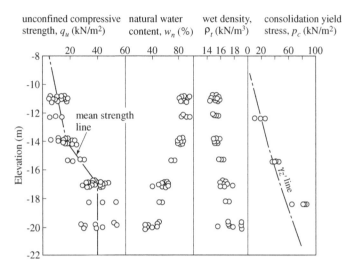

Figure 5.6.   Soil condition at test site [3,6].

Figure 5.6 shows the soil condition at the test site, which consists of a bearing stratum of mudstone, sand and gravel layer of 0 to 2 m in thickness, and alluvial clay layer of about 10 m in thickness above the sand and gravel layer. The unconfined compressive strength, q_u, of the soft clay layer increases to about 40 kN/m² at -17 m. The natural water content, w_n, of the layer decreases almost linearly with depth. The measured data on the shear strength, water content, wet density and consolidation yield stress distribution with depth clearly show that the soft alluvial clay layer was in the normally consolidated condition.

Figure 5.7 shows the time schedule of the loading test, where the sand piles installation and field loading test as well as very detailed soil surveys were conducted within two years [3,6]. After the sand piles installation, the soil characteristics were measured to investigate the shear strength reduction and recovery due to the sand piles installation. The consolidation behavior of the improved ground including the stress concentration was measured during the placement of superstructure. The loading test was carried out for 10 months after the sand piles installation.

The test conditions are illustrated in Figure 5.8, which consists of a SRC slab, three concrete caissons and three water tanks on the SCP improved ground [3,6]. After spreading the sand blanket (1) of about 1 m in thickness on the soft ground, sand piles of 1.7 m in diameter were constructed at an interval of 3.0 m, which corresponded to the replacement area ratio, as, of 0.35. Other sand piles of 2.0 m in diameter were constructed in the side area at an interval of 2.1 m, as of 0.7, to ensure that shear failure at the loading stage would take place in the designed direction, which was necessary in order to reduce the instrumentation cost. The characteristics of the sand piles were surveyed by the standard penetration tests.

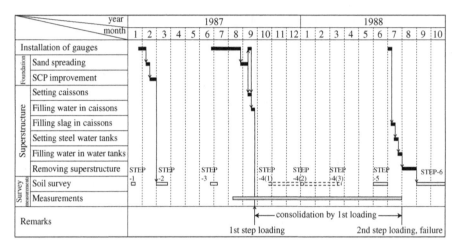

Figure 5.7.   Schedule of field test [3,6].

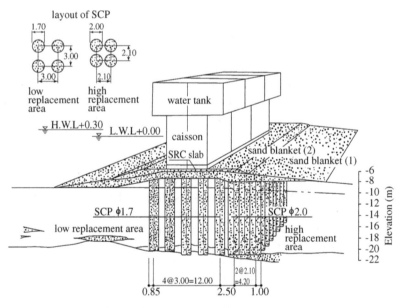

Figure 5.8.   Illustration of test site [3,6].

Many transducers including earth pressure cells, piezometers, inclinometers and load cells were installed in the ground before and after the improvement, as shown in Figure 5.9 [3,6]. After the construction of sand piles, the sand blanket (2) and the SRC slab of 7.5 m in width by 30.2 m in length by 0.7 m in thickness were placed on the

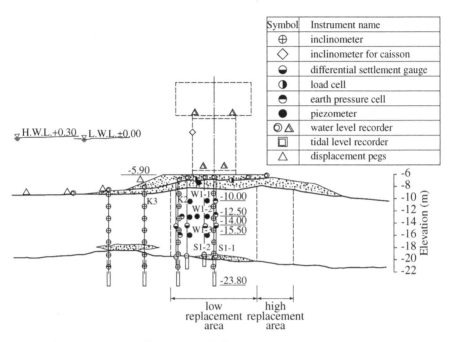

Symbol	Instrument name
⊕	inclinometer
◇	inclinometer for caisson
◒	differential settlement gauge
◑	load cell
◓	earth pressure cell
●	piezometer
◐△	water level recorder
▢	tidal level recorder
△	displacement pegs

Figure 5.9. Arrangement of instruments [3,6].

improved ground. Next, three concrete caissons were placed on the slab, which corresponded to a total effective vertical stress of $30\,kN/m^2$. The improved ground was
allowed to be consolidated by this vertical pressure for about 10 months. During the
consolidation, the settlement of caissons, the pore water pressure, and the earth pressure at many locations in and around the improved ground area were measured to
investigate the consolidation settlement, the reduction and recovery of shear strength
of clay, the stress concentration and so on. After the completion of the ground consolidation, the second stage loading was started on July 17, 1988 by filling slag into the
concrete caissons by which the total vertical stress reached approximately $80\,kN/m^2$.
Three steel water tanks were placed on the top of the caissons for the following eight
days. During this period, the total vertical stress was kept almost constant by cancel-
ing the weight of the steel tanks by dewatering in the caisson. On July 26, the final
loading was carried out by supplying water into the concrete caissons and the steel
tanks until the ground completely failed.

Figure 5.10 shows the change in shear strength of the clay in and around the
improved ground, which were measured before and after the sand piles installation
[3,6]. In the figure, the unconfined compressive strength distributions just after the
sand piles installation, four months after installation and just before the second loading stage (about 10 months after installation) are plotted with depth, together with
those before the improvement, as a chained line. The figure clearly shows that the

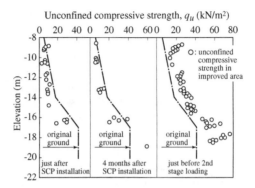

Figure 5.10.   Unconfined compressive strength change after SCP installation [3,6].

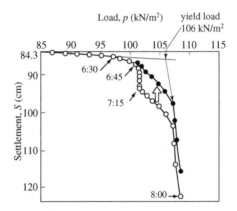

Figure 5.11.   Relationship between load and ground settlement [3,6].

unconfined compressive strength just after the sand piles installation was smaller than that of the original ground, however, it recovered quickly to 90% of the original strength at four months after installation. At ten months after, just before the second loading stage, the unconfined compressive strength became 80% larger than that of the original ground throughout the depth which was due to the pore water pressure dissipation generated during the sand piles installation and the vertical stress increase of the concrete caissons.

Figure 5.11 shows the relationship between the vertical load and the settlement of caissons, which was measured during the second loading stage and plotted as open circles [3,6]. At 6:45 on July 18, the vertical load was kept constant for about 15 minutes in order to change the pipeline of the water supply to the steel water tanks. The settlement of caissons took place continuously during this period. Thereafter, the ground settlement increased very rapidly as the vertical load was increased, and the ground finally failed with large rotational movement of caissons as shown in Figure 5.12 [8].

Figure 5.12.   A series of photographs during the final loading [8].

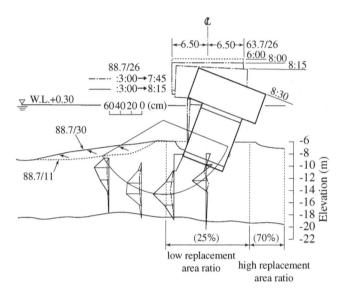

Figure 5.13.   Ground deformation of improved ground [3,6].

A part of the load – settlement curve is shifted as full circles to compensate for the settlement during changing the pipeline. The yield load was obtained as the load at the intersection of the initial and final straight lines of the load – settlement curve as shown in the figure, and was 106 kN/m².

After the completion of the loading test, the water tanks and concrete caissons for the loading test were removed and the improved ground was surveyed throughout to detect the failure plane and the shape of ground heaving due to the failure. Figure 5.13 shows the horizontal displacement in the ground and the displacement of the caisson due to the vertical loading [3,6]. The figure clearly indicates that a circle shaped ground failure took place together with large rotational movement of the caissons. Figure 5.14 illustrates the results of laboratory tests and field cone penetration tests performed after the loading test [3,6]. In the laboratory tests, unconfined compression tests were conducted on the sampled specimen. The black marked portion in the figure shows the area where the soil strength measured by the laboratory and field tests was much lower than that of the surrounding area. This means that the slip surface passed through this area. In the figure, the failure plane by the slip circle analysis is also plotted, which coincides well with the test results. The figure also confirms that a circle shaped ground failure took place.

The measured bearing capacity of the improved ground was compared with a slip circle analysis with Boussinesq's stress distribution method in which the shear strength formula for the improved ground was expressed as Equation (2.16a) and the shear strength increment due to the vertical pressure in the first loading stage was taken into account. The safety factor was calculated by two-dimensional and

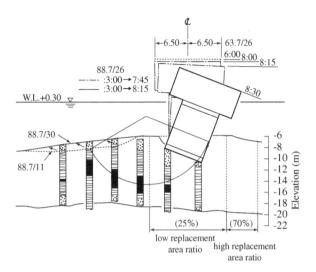

Figure 5.14.   Sliding surface estimated by shear strength [3,6].

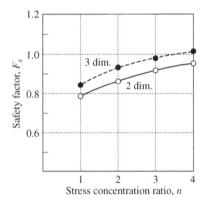

Figure 5.15.   Relationship between calculated safety factor and stress concentration ratio [3].

three-dimensional slip circle analyses. In the three-dimensional slip circle analysis, the shear strength mobilizing on the side faces of the slip circle was taken into account [9]. Figure 5.15 shows the relationships between the calculated safety factor, Fs, and the stress concentration ratio, n [3]. The safety factor increases gradually with increasing stress concentration ratio irrespective of the type of slip circle analysis. The safety factors calculated with n of 3 were 0.92 and 0.98 for two-dimensional and three-dimensional analyses respectively. These values were quite reasonable for evaluating the actual bearing capacity. These confirmed that the bearing capacity of the improved ground with a low replacement area ratio could be evaluated accurately by the slip circle analysis with Equation (2.16a) as a shear strength formula and Boussinesq's stress distribution method.

This field test revealed various characteristics of SCP improved ground and helped to establish the SCP design method with a low replacement area ratio for port facilities [10–11]. This test also helped to construct the sea revetments of the man-made island in the Kansai International Airport Construction Project more economically.

5.4   HORIZONTAL RESISTANCE OF PILE STRUCTURES

The Trans-Tokyo Bay Highway Project was planned to improve heavy traffic congestions at Metropolitan Tokyo area by constructing a by-pass between Kanagawa and Chiba Prefectures (Figure 5.16). In this project, a new 15.1 km-long highway as well as two man-made islands were constructed in the Tokyo Bay, as shown in Figure 5.17 [12–13]. The SCP method was applied at Kawasaki man-made island and Ukishima access. Kawasaki man-made island as shown in Figures 5.18a and 5.18b was constructed for starting point of shield tunnel machines to Kisarazu man-made island and Ukishima access during the construction, and for a ventilation tower after the construction [13–14]. The island was a sort of structural island, which consisted of a soil mixing wall with a diameter of about 100 m and outer and inner steel jackets. The SCP method was applied to assure the horizontal resistance of the inner and outer jackets and to improve the trafficability at the center part of island during excavation. The center part of the island was excavated to the depth of -69.7 m together with removing the inner jacket structure for setting the shield tunnel machines.

Figure 5.19 shows the ground condition at Kawasaki man-made Island, where the water depth was 28 m [13]. An alluvial clay layer (Ac_1 layer) was deposited with a

Figure 5.16.   Trans-Tokyo Bay Highway Project.

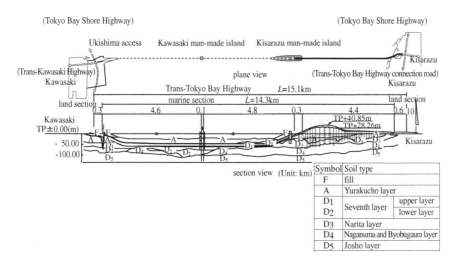

Figure 5.17.   Sectional view of Trans-Tokyo Bay Highway [13].

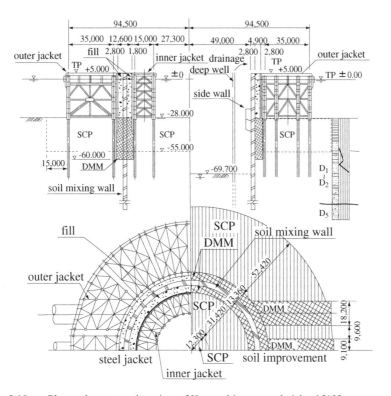

Figure 5.18a.   Plan and cross section view of Kawasaki man-made island [13].

Figure 5.18b.   Photograph of Kawasaki man-made island [14].

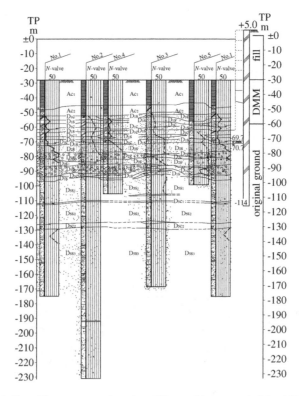

Figure 5.19.   Soil profile at construction site of Kawasaki man-made island [13].

thickness of about 27 m on the sea bed, whose SPT N-value was around zero. Several diluvial clay and sandy layers were stratified alternately beneath the alluvial clay layer, whose total thickness was about 35 m. The diluvial sandy layer had a high SPT N-value of around 10 to 70. The stiff sandy layer with SPT N-value exceeding 70 was found at about -90 m.

Figure 5.18a also shows the sectional view of the soil improvement. The ground in the island was improved by the SCP method and the Deep Mixing method up to the depth of -55 m to assure the horizontal stability of the steel jacket type structure and the stability of the excavated face of the soil mixing wall. The SCP method was adopted at the area of the jacket structure and at the center part of the island. At the former area, sand piles of 1.6 m in diameter were constructed with the replacement area ratio of 0.785 in order to assure the horizontal resistance of the jacket piles. At the center part of the island, sand piles of 1.4 m in diameter were constructed with the replacement area ratio of 0.301 in order to assure trafficability during excavation. The sand piles at both areas were constructed by the vertical vibrating compaction technique and the vertical and horizontal vibrating compaction technique (see Chapter 6.4). The construction was successfully completed within four months in which four SCP barges having three to six casing pipes were operated as shown in Figure 5.20 [14].

This construction site was of interest, so the constructed sand piles under the sea bed were investigated in detail during the excavation. A detailed field investigation of the sand piles was carried out about four years after the completion of SCP

Figure 5.20.   Lineup of SCP barges at construction site [14].

Figure 5.21a.   Photographs of sand pile heads during excavation at the replacement area ratio of 0.301 [13].

Figure 5.21b.   Photographs of sand pile heads during excavation at the replacement area ratio of 0.785 [13].

construction. The investigated area was improved by the vertical vibrating compaction technique. Figures 5.21a and 5.21b show photographs of sand pile heads at replacement area ratios of 0.301 and 0.785 respectively [13]. Figure 5.22 shows the water content and unconfined compressive strength profiles of the original ground, the sand pile ratio and the SPT N-value distribution with depth [13]. The figure shows the unconfined compressive strength, q_u, of the original ground is quite small at the surface and increases almost linearly with depth. The SPT N-values shown in the figure

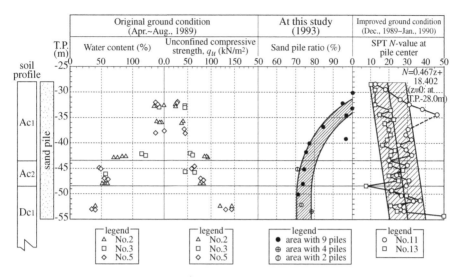

Figure 5.22.   SPT N-value and sand pile ratio with depth [13].

were measured at the completion of improvement. Although there is much scatter in the measured data, the SPT N-values show relatively high values exceeding about 10 to 40 and increase with depth. During the excavation, the sand pile ratio was measured at every couple of meters depth up to the final excavation depth of -55 m. The sand pile ratio is defined as the ratio of the sectional area of sand piles to total sectional area of improved ground, which is the same definition as that of the replacement area ratio. The measured data is also shown in Figure 5.22 [13]. This figure shows that the sand pile ratio at large depths is about 72% to 75%, which almost corresponds to the design value, but increases by more than 90% at shallow depths. This means that the diameter of compacted sand piles becomes larger at shallow depths. This phenomenon can be explained by the fact that a larger amount of sand was necessary at the shallow area to assure a sufficient degree of compaction where the ground strength and confined pressure were small.

A detailed soil survey on the compacted sand piles was also conducted during the excavation; the fines content distribution of the compacted sand piles is shown in Figure 5.23 [13]. The fines content of the sand piles, Fc, is almost less than 5% except at the depth of around -45 m. This means that there is no possibility of clay particles intruding into the sand piles even four years after sand piles installation.

The characteristics of the clay ground between the sand piles were also investigated during the excavation. Figure 5.24 shows the soil properties of the clay ground between the sand piles, in which the water content, w_t, wet density, ρ_t, and unconfined compressive strength, q_u, are plotted with depth [13]. It is found that the water content of the clay decreases greatly from the original value. This phenomenon is clearly found in the Ac_1 layer after four years. The ground strength increased by about

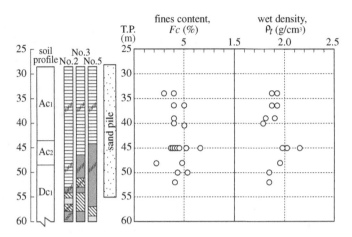

Figure 5.23.   Physical properties of sand piles [13].

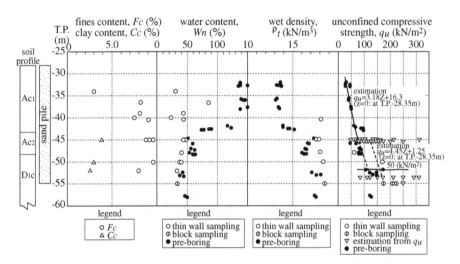

Figure 5.24.   Properties of clay ground between sand piles [13].

$50 \, \text{kN/m}^2$ in unconfined compressive strength even if no overburden pressure was applied on the ground, which was caused by the dissipation of the excess pore water pressure generated during the sand piles installation. The shear strength of the Ac_1 layer was not measured. However, the strength increase is expected to be relatively large in view of the large decrease in water content.

   In conclusion, the sand piles constructed in the undersea clay ground remained in good condition in terms of shape, quality and performance.

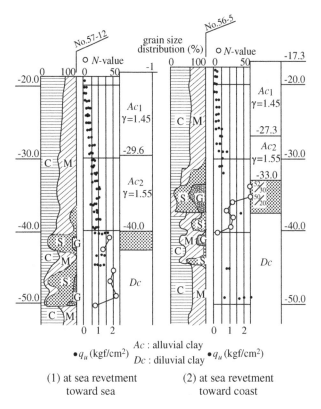

Figure 5.25.   Typical ground condition [15].

5.5   STABILITY OF STEEL CELL TYPE QUAY

A man-made island for Kansai International Airport was constructed 5 km from the shoreline at Osaka Bay. The island for the first construction phase of the project had an area of 5.1 million m^2 and was reclaimed with about 150 million m^3 of mountainous soils on the soft ground. The first construction phase of the airport was started in 1987 with the construction of sea revetments, sea reclamation, airport facilities and a connecting bridge, and was completed in 1994, and another man-made island is now being constructed as the second phase to expand the capacity of the airport.

Figure 5.25 shows a typical ground condition at the site where two soft alluvial clay layers, Ac_1 and Ac_2, were stratified to the depth of $-33\,m$ to $-40\,m$ [15]. The unconfined compressive strength, q_u, of the layers increased almost linearly with depth, indicating that the alluvial clay layers were in a normally consolidated condition. Several sandy layers and diluvial clay layers were stratified alternately underneath the alluvial clay layers to a depth of a couple of hundred meters.

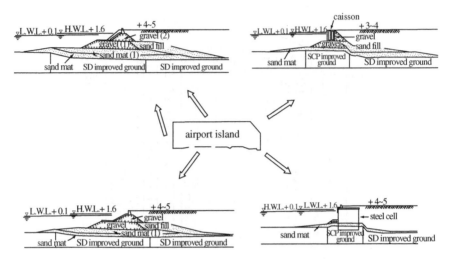

Figure 5.26.   Illustration of sectional area of sea revetment of man-made island [15].

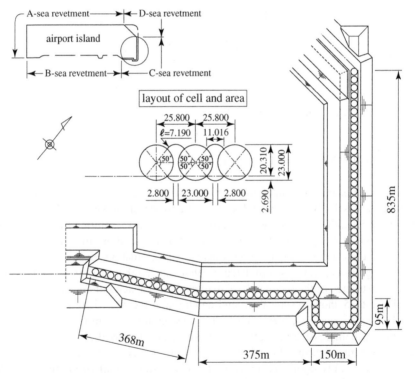

Figure 5.27.   Layout of sea revetment to be improved by sand compaction pile method [16].

The sea revetments of total length of about 11 km were classified into roughly four structural types as shown in Figure 5.26 [15]. The Sand Compaction Pile method was adopted at two of the four sea revetments, the caisson type and the cellular type, to assure the stability of the revetments.

The construction site introduced here is the steel cellular type sea revetment (C-sea revetment) at the corner of the island, as shown in Figures 5.27 and 5.28 [16]. After spreading a sand mat with a thickness of about 3 m, compacted sand piles of 2 m in diameter were constructed in a square pattern of 2.1 m intervals (Figure 5.28), which corresponded to the replacement area ratio of 0.712. Figure 5.29 shows the lineup of SCP barges at the construction site [14]. The sand material used here was gravel with relatively large particle size instead of ordinary marine sand. During the sand piles installation, a large amount of ground upheaval took place. Based on the field test results at Maizuru Port as introduced before, the upheaval soil was not excavated but was also improved by the method to assure the drainage function of the sand piles. In this portion, sand piles without any compaction were constructed in a similar manner

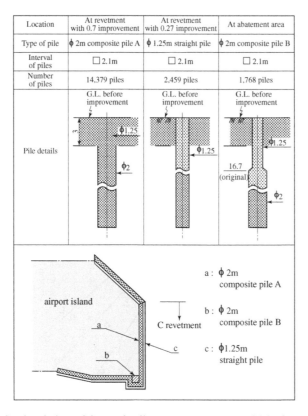

Figure 5.28.   Sectional view of the steel cell type sea revetment and SCP improvement [16].

to the Sand Drain method. The diameter of the sand piles was 1.25 m, which was the same as the outer diameter of the casing pipe (see Figure 5.28). This was essential to assure sufficient hydraulic connection of the sand piles and the sand mat. After the installation of sand piles, steel cells of 23 m in diameter were penetrated into the

Figure 5.29.   Lineup of SCP barges at construction site [14].

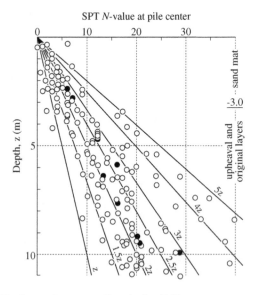

Figure 5.30.   SPT N-value at the center of sand piles [16].

improved ground to the depth of −19 m by means of eight sets of vibro-hammers, which was soon followed by filling sand in the cells.

The SPT N-values at the center of sand piles are shown in Figure 5.30 with depth [16]. These data were measured at the upper 11 m layer of the improved ground where the steel cells were installed. The figure shows that the SPT N-value increases with depth with a large scatter. The reason for the scatter is thought to be the effect of large particle size of the gravel used as a SCP material.

5.6   COPPER SLAG AS SCP MATERIAL

More than 1.8 million tons of copper slag, a by-product of the refining process of copper, are produced annually in Japan [17]. Since the slag has relatively large specific gravity of soil particles and high permeability, some amount of the slag has been used for a sand mat material and a sand material filled within concrete caissons in port and harbor constructions. However, large amounts of the slag have been dumped in disposal areas. Recently, the limited availability of disposal areas has necessitated research to find new applications for the slag, such as for a SCP material. To investigate the workability and construction ability of SCP improvement with copper slag, a field construction test was performed at Uno Port in Okayama Prefecture. At Uno Port, a sheet pile wall type revetment was planned to be constructed at the front side of the existing caisson type revetment for renovating port and harbor facilities. In the design, the clay grounds at the front and rear sides of the sheet pile wall were improved by the Sand Compaction Pile method with the replacement area ratio of 0.7, as shown in Figure 5.31 [18]. The SCP improvement was originally designed to be carried out with marine sand only. However, for this field test, the design was slightly changed by dividing the improved area into two areas for SCP, one with copper slag and the other with marine sand.

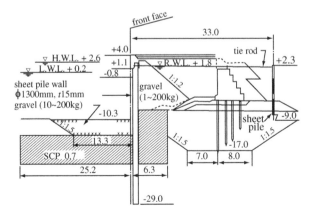

Figure 5.31.   Sectional view of construction site [18].

Table 5.3.   Physical properties of copper slag and marine sands [18].

			Particle size distribution					Unit weight (kN/m³)	
		Gs	max. (mm)	fines (%)	sand (%)	gravel (%)	U_c (%)	max.	min.
Copper slag	un-collapsed	3.511	4.75	0.0	79.0	21.0	2.2	19.9	16.2
	collapsed	3.511	4.75	2.0	91.0	7.0	2.5	21.7	17.4
Marine sand	Ehime	2.647	9.52	0.3	92.4	7.3	4.3	–	–
	Kagawa	–	–	0.0	81.0	19.0	4.0	–	–

The copper slag was a black, granular material. Its physical properties are summarized in Table 5.3 together with those of the marine sand, which were used for the field construction test [18]. It is also found that the unit weights of the copper slag are larger than those of the marine sand.

A typical soil condition of the site is shown in Figure 5.32 [18]. It can be seen that a soft alluvial clay layer, Ac, was deposited from DL −10 m to −20 m overlaying a diluvial sandy layer, Ds, of about 5 m in thickness. The diluvial sandy and clay layers, Ds and Dc, were deposited alternately below DL −20 m. The unconfined compressive strength, q_u, of the Ac layer increased with depth to about 100 kN/m² at DL −20 m, which showed that the alluvial clay layer was in a normally consolidated condition. The SPT N-values are also plotted in Figure 5.32. It can be seen that the SPT N-values are larger values of around 20 to 50 in the Ds layer and smaller values of about 10 in the Dc layer. Thus, the soft alluvial Ac layer of DL −10 m to −20 m was improved by the Sand Compaction Pile method.

SCP improvement was conducted during about two weeks in 1995 by the vertical vibrating compaction technique as explained later in Chapter 6.4. The SCP machine used in the field construction test was an ordinary machine for marine sand, and was also used for the copper slag without any machinery changes.

After the SCP execution, check borings were performed in order to confirm the strengths of compacted sand piles and copper slag piles. The SPT N-values measured in the standard penetration test are plotted with depth in Figure 5.33 in which the test data of copper slag piles are plotted together with those of the marine sand piles [18]. Although the test data has a large scatter, it is found that the SPT N-values of both piles increase almost linearly with depth, and that the SPT N-values of the copper slag piles are slightly larger than those of the marine sand piles. Grain size analyses were performed on the samples, and the test results are summarized in Figure 5.34 [18]. It can be seen in the figure that the particle size distribution curve after the construction is slightly larger than that before, which is probably because some amount of slag particles collapsed due to the compaction *in situ*. However, it can be concluded that these small changes in

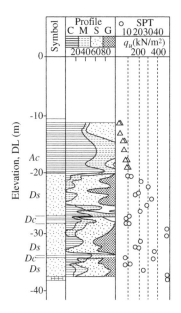

Figure 5.32.   Soil condition of the site [18].

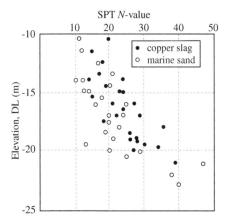

Figure 5.33.   SPT *N*-value with depth [18].

particle size distribution have negligible effect on the strength and permeability of compacted piles.

From the field test, the SCP improved ground with the copper slag has relatively large shear strength, of the same order as that of ordinary marine sand, and hence copper slag is applicable for the Sand Compaction Pile method.

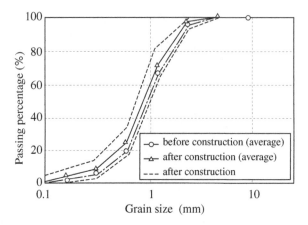

Figure 5.34.   Sand particle analysis [18].

5.7   OYSTER SHELL AS A SCP MATERIAL

Oyster is cultured in several areas in Japan, especially in Hiroshima and Miyagi Prefectures, with about 4,000 tons of oyster cultured in Miyagi Prefecture every year. As a result, 30 tons of oyster shell are produced every year. About a quarter of the oyster shell has been used beneficially, e.g. for agriculture. However, about three quarters have been disposed of or left unused, which causes environmental problems such as foul odor in the surrounding area. In view of the circumstances, a field test was carried out at Ishinomaki Port in Miyagi Prefecture to investigate the applicability of oyster shell as a sand compaction pile material [19–20].

Figure 5.35 shows a sectional view of SCP improved ground and a breakwater, in parts of which a mixture of oyster shell and sand (25% : 75% in volume) was used as a SCP material [20]. Figure 5.36 shows the particle size distribution of the soil and the oyster shell [20]. Sand piles with a diameter of 1.96 m were constructed at an interval of 2 m, which corresponded to the replacement area ratio of 0.754. After the execution, standard penetration tests were conducted at the oyster and sand mixture piles as well as the sand piles. The relative density and the internal friction angle of mixture piles were estimated from the measured SPT N-values, and are plotted in Figure 5.37 with depth [20]. The relative density of the shell and sand mixture pile was quite high, exceeding 100%, which could be explained by the difference in compaction technique between the laboratory tests and the field execution, and by the effect of crushing oyster shells. This indicates that the shell and sand mixture was well compacted *in situ*. The internal friction angle estimated by the relative density and direct shear tests shows a high value exceeding 40 degrees. These data have clearly revealed the high applicability of oyster shell as a SCP material.

Section

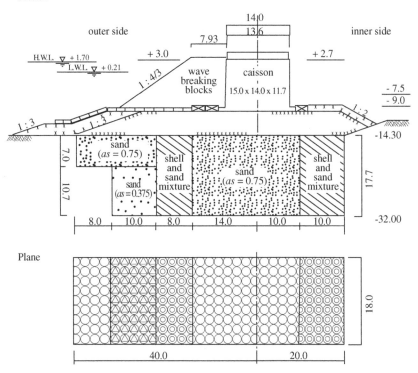

Plane

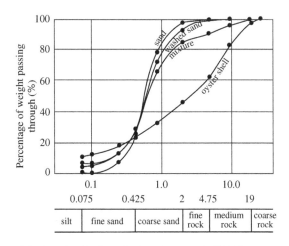

Figure 5.35.   Sectional area of foundation and breakwater at Ishinomaki Port [20].

Figure 5.36.   Particle size distribution of soil and oyster shell [20].

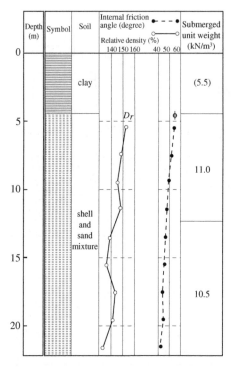

Figure 5.37.   Soil profile of mixture of oyster shell and sand [20].

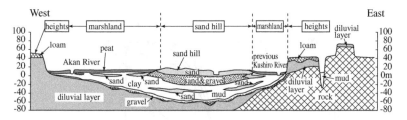

Figure 5.38.   Soil profile at Kushiro area [21].

5.8   LIQUEFACTION PREVENTION

At Kushiro, which is in the far east of Hokkaido island, thick marshlands composed of soft clay, peat and mud deposits are widely deposited as shown in Figure 5.38 [21]. Furthermore, sand layers under the peat and sand hills are very loose and saturated, which means that there is a high likelihood of liquefaction during an earthquake. Sea reclamations have been frequently carried out at coastal areas of Kushiro Port for

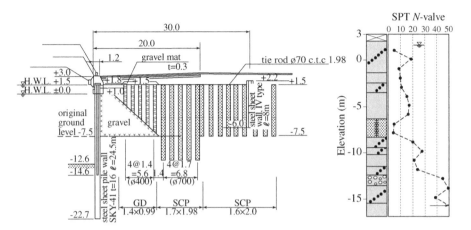

Figure 5.39.   Sectional view of sheet pile wall quay at Kushiro West Port [22].

creating port and industrial areas. In the Kushiro West Port area, almost all sea recla-mations were constructed by using sandy soil dredged there.

Figure 5.39 shows a sectional view of the steel sheet pile wall type quay at Kushiro West Port, which was back filled with gravel and sandy soils [22]. Figure 5.39 also shows the soil profile of the back filled sandy soil, in which loose sandy layers with the SPT N-value of around 10 were deposited to the depth of about -8 m. As the sandy layer was loose and saturated, liquefaction was anticipated at the site during an earth-quake. In order to prevent liquefaction, the loose sandy layer was improved by the Sand Compaction Pile (SCP) method and the Gravel Drain (GD) method in 1989, as shown in Figure 5.39. The Sand Compaction Pile method was conducted up to the depth of -7.5 m or -12.0 m. Sand piles of 70 cm in diameter were installed at an interval of 1.7 m, which corresponded to the replacement area ratio, as, of 0.133. The Gravel Drain method was applied at the area close to the steel sheet pile wall to prevent adverse influ-ences on the steel pile wall structure due to the sand piles installation, whose diameter and interval were 0.4 m and 1.4 m respectively.

In 1993, a large earthquake with a magnitude of 7.8 struck in the Kushiro area and caused heavy damage (Kushiro-oki earthquake). At Kushiro Port, some sea revet-ments and quay walls without any soil improvement were heavily damaged with large ground cracks as shown in Figure 5.40 [21]. Figure 5.41 shows a photograph of a quay wall at the improved area in Kushiro Port after the Kushiro-oki earthquake [21]. The figure shows that there was no damage at the site, and loading and unloading works could continue without any inconvenience the day after the earthquake.

Figure 5.42 shows another example of SCP improvement effect where oil tank foundations were improved by the SCP method with the replacement area ratio of 0.115 in 1977 [21]. Figure 5.43 shows the SPT N-values before and after the improvement [21]. The SPT N-value after the improvement was around 15 to 40,

Figure 5.40.   Damage of quay wall at Kushiro Port caused by Kushiro-oki earthquake [21].

Figure 5.41.   Photograph of quay wall at the improved ground area at Kushiro Port after Kushiro-oki earthquake [21].

whereas the original value was less than 10. No damage was caused at the oil tanks and foundation by the earthquake in 1993.

5.9   STATIC COMPACTION TECHNIQUE

The Sand Compaction Pile method was adopted for coastal embankment as a countermeasure to prevent liquefaction at Matsusaka Port in Mie Prefecture. The ground condition at the construction site was as shown in Figure 5.44, where a loose sand

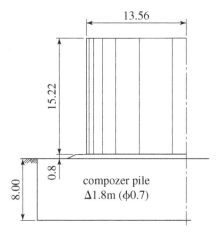

13.56

15.22

0.8

8.00

compozer pile
Δ1.8m (φ0.7)

Figure 5.42.   Sectional view of oil tank and foundation at Kushiro West Port [21].

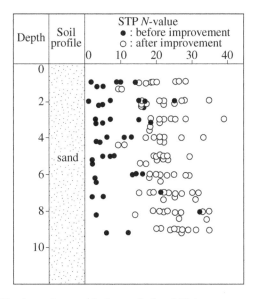

Figure 5.43.   SPT N-values of ground before and after SCP improvement [21].

layer with the SPT N-value of less than 15 was deposited [23]. The sand layer was con-sidered likely to be liquefied during an earthquake according to its loose and saturated condition. As a countermeasure against liquefaction, sand piles with a diameter of 70 cm were constructed to the depth of -12.5 m in a square pattern with an interval of 1.6 m, which corresponded to the replacement area ratio, a_s, of 0.15 (Figure 5.45

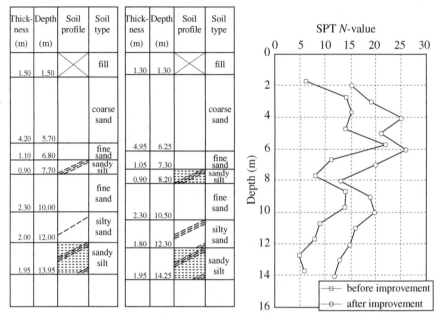

Thick-ness (m)	Depth (m)	Soil profile	Soil type		Thick-ness (m)	Depth (m)	Soil profile	Soil type
1.50	1.50		fill		1.30	1.30		fill
4.20	5.70		coarse sand		4.95	6.25		coarse sand
1.10	6.80		fine sand		1.05	7.30		fine sand
0.90	7.70		sandy silt		0.90	8.20		sandy silt
2.30	10.00		fine sand		2.30	10.50		fine sand
2.00	12.00		silty sand		1.80	12.30		silty sand
1.95	13.95		sandy silt		1.95	14.25		sandy silt

Figure 5.44.   Soil profile at site [23].

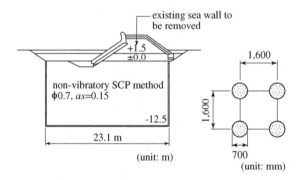

Figure 5.45.   Cross section of improved ground and sand pile arrangement [24–25].

[24–25]). Since there were many residential houses close to the construction site, a static compaction technique (non-vibrating compaction technique) was adopted instead of a vibrating compaction technique in order to prevent adverse environmental impacts such as vibration and noise to the surrounding area. The vibration and noise were measured at several points, 10 m, 30 m, 50 m, 100 m and 500 m distant from the SCP machine during the construction to ensure negligible adverse impact on the environment. Figures 5.46a and 5.46b show the measured data on the vibration

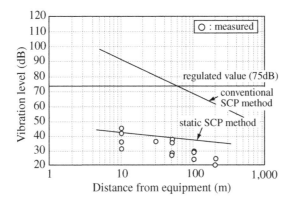

Figure 5.46a.   Vibration measurements during execution [24].

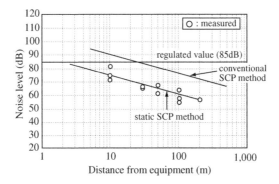

Figure 5.46b.   Noise measurements during execution [24].

and noise with distance respectively [24]. In the figure, the data by a vibrating compaction technique are also plotted for comparison, which are cited from previous field experiences. It can be seen that those generated by the static compaction technique are smaller than the vibrating compaction technique, and are lower than the regulated levels set by Ministry of the Environment, Japan.

Figure 5.47 shows the relationship between the SPT N-value of the ground between the sand piles and the fines content, Fc, of the original ground [24–25]. The SPT N-value plotted in the figure is the SPT N-value, N_1, normalized by Equation (5.1) [26]. The figure shows that the SPT N-value after the improvement increases to 10 to 30 from the original value of 5 to 15, while it decreases with increase of fines content.

$$N_1 = \frac{170 \cdot N}{\sigma'_v + 70}$$

(5.1)

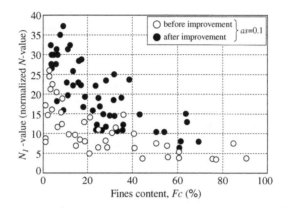

Figure 5.47.   Relationship between normalized SPT N-values before and after improvement and fines content [24–25].

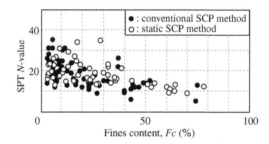

Figure 5.48.   Relationship between normalized SPT N-value and fines content [27].

where:

N : measured SPT N-value

N_1 : normalized SPT N-value

σ'_v : overburden pressure at measured depth (kN/m²)

At the site, SCP improvement with the vibrating compaction technique had been conducted before. Figure 5.48 compares the normalized SPT N-values after the improvement by these two techniques [27]. Although there is much scatter in the measured data, the static compaction technique adopted at the site has a similar improvement effect to the vibrating compaction technique.

REFERENCES

1   Kurihara, N. and Tochigi, H.: Consideration on the deformation characteristics of sand compaction pile method with field measurement. Proc. of the Symposium

on Strength and Deformation of Composite Ground, pp. 135–138, 1984 (in Japanese).

2 Suematsu, N., Isoda, T. and Kanda, Y.: Construction of highway on soft ground. Proc. of the International Symposium on Embankment on soft ground, – Test embankment trial –, 1985.

3 Okada, Y., Yagyuu, T. and Sawada, Y.: Field loading test of SCP improved ground with low replacement area ratio. Journal of the Japanese Society of Soil Mechanics and Foundation Engineering, 'Tsuchi-to-Kiso', Vol.37, No.8, pp. 57–62, 1989 (in Japanese).

4 Terashi, M., Kitazume, M. and Minagawa, S.: Bearing capacity of improved ground by sand compaction piles. Deep Foundation Improvements: Design, Construction, and Testing, ASTM STP 1089, Robert C. Bachus, Ed., American Society for Testing and Materials, pp. 47–61, 1990.

5 Terashi, M. and Kitazume, M.: Bearing capacity of clay ground improved by sand compaction piles of low replacement area ratio. Report of the Port and Harbour Research Institute, Vol. 29, No. 2, pp. 119–148, 1990 (in Japanese).

6 Okada, Y.: Ground deformation and strength of improved ground by SCP method with low replacement area ratio. Ph.D. Theses, 1993 (in Japanese).

7 Asaoka, A., Kodaka, T. and Nozu, M.: Undrained shear strength of clay improved with sand compaction piles. SOILS AND FOUNDATIONS, Vol. 34, No. 4, pp. 23–32, 1994.

8 Coastal Development Institute of Technology: Brochure.

9 Nakase, A. and Kobayashi, M.: Bearing capacity of foundation on cohesive soil under eccentric and inclined loads. Report of the Port and Harbour Research Institute, Vol. 9, no. 2, pp. 23–38, 1970 (in Japanese).

10 Ministry of Transport: Technical standards and commentaries for port and harbour facilities in Japan. Ministry of Transport, Japan, 1999 (in Japanese).

11 The Overseas Coastal Area Development Institute of Japan: English version of technical standards and commentaries for port and harbour facilities in Japan. 2002.

12 Uchida, K., Shioji, Y. and Kawase, Y.: Cement treated soil in the Trans-Tokyo Bay Highway project. Proc. of the 13th International Conference on Soil Mechanics and Foundation Engineering, pp. 1179–1182, 1993 (in Japanese).

13 Kobayashi, H., Kogo, M., Suzuki, K. and Sakai, S.: Estimation of the clay ground improved by sand compaction piles at Kawasaki man-made island. Journal of Construction Management and Engineering, Japan Society of Civil Engineers, No. 553/6-33, pp. 41–48, 1996 (in Japanese).

14 The courtesy of Fudo Construction Co. Ltd.

15 Furudoi, T. and Yajima, M.: Outline of Kansai International Airport construction. Journal of the Japanese Society of Soil Mechanics and Foundation Engineering, 'Tsuchi-to-Kiso', Vol. 34, No. 1, pp. 13–18, 1986 (in Japanese).

16 Takai, T., Imano, K., Ogino, H. and Nakamura, M.: A study of cellular shell driving into ground using vibration hammers. Journal of Construction Management

and Engineering, Japan Society of Civil Engineers, No. 415/6-12, pp. 53–62, 1990 (in Japanese).

17   Minami, K., Matsui, H., Naruse, E. and Kitazume, M.: Field test on sand compaction pile method with copper slag sand. Journal of Construction Management and Engineering, Japan Society of Civil Engineers, No. 574/6-36, No. 574/6-36, pp. 49–55, 1997 (in Japanese).

18   Kitazume, M., Minami, K., Matsui, H. and Naruse E.: Field test on applicability of copper slag sand to sand compaction pile method. Proc. of the 3rd International Congress on Environmental, Vol .2, pp. 643–648, 1988.

19   Hashidate, Y., Fukuda, T., Okumura, T. and Kobayashi, M.: Engineering properties of oyster shell-sand mixtures. Proc. of the 28th Annual Conference of the Japanese Society of Soil Mechanics and Foundation Engineering, pp. 869–872, 1993 (in Japanese).

20   Hashidate, Y., Fukuda, T., Okumura, T. and Kobayashi, M.: Engineering properties of oyster shell-sand mixtures and their application to sand compaction piles. Proc. of the 29th Annual Conference of the Japanese Society of Soil Mechanics and Foundation Engineering, pp. 717–720, 1994 (in Japanese).

21   Fudo Construction Co., Ltd.: Report of disasters by Kushiro-Oki earthquake. 1993 (in Japanese).

22   Ando, Y., Tsuboi, H., Yamamoto, M., Harada, K. and Nozu, M.: Recent soil improvement methods for preventing liquefaction. Proc. of the 1st International Conference on Earthquake Geotechnical Engineering, pp. 1–6, 1995.

23   Waga, A., Asami, Y., Fukada, H., Nakai, N., Harada, K. and Nozu, M.: A consideration about improvement area of non-vibratory sand compaction method (Save compozer). Proc. of the 33rd Annual Conference of the Japanese Society of Soil Mechanics and Foundation Engineering, pp. 2159–2160, 1998 (in Japanese).

24   Tsuboi, H., Ando, Y., Harada, K., Ohbayashi, J. and Matsui, T.: Development and application of non-vibratory sand compaction pile method. Proc. of the 8th International Offshore and Polar Engineering Conference, pp. 615–620, 1998.

25   Yamamoto, M. and Nozu, M.: Effects on environmental aspect of new sand compaction pile method for soft soil. Proc. of Coastal Geotechnical Engineering in Practice, pp. 563–568, 2000.

26   Tokimatsu, K. and Yoshimi, Y.: Empirical correlation of soil liquefaction based on SPT N-value and fines content. SOILS AND FOUNDATIONS, Vol. 23, No. 4, pp. 56–74, 1983.

27   Suganuma, M., Fukada, H. and Nakai, N.: Case history of non-vibratory sand compaction method. Proc. of the 52nd Annual Conference of the Japan Society of Civil Engineers, III, pp. 412–413, 1997 (in Japanese).

Chapter 6

Development of SCP Machines and Current Techniques

6.1 INTRODUCTION

Several types of SCP machine have been developed and adopted for on-land construction and marine construction for more than 50 years, which include hammering compaction technique, vibrating compaction technique and static compaction technique (non-vibrating compaction technique). In this section, the development of SCP machines with various techniques is introduced in detail as well as the current techniques in Japan.

6.2 MACHINERY DEVELOPMENT

The principle of the Sand Compaction Pile method was introduced in the 1950s. In 1956, the first type of SCP machine as shown in Figure 6.1 was developed in Japan for densification of sandy ground. In Figure 6.1a, a casing pipe with a hammering probe was suspended along a wooden scaffold. The SCP machine in Figure 6.1b, on the other hand, consisted of a crawler crane, a casing pipe and a hammering probe. The casing pipe of 30 cm in diameter was penetrated into the ground, then sand piles were constructed by feeding sand material during the stage of retrieving the casing pipe. In this machine, sand fed out from the casing pipe was compacted by a similar manner to the pile driving technique, by dropping a heavy weight (ram) onto the sand in the casing pipe (hammering compaction technique). This method was called as the 'Compozer Method' in those days.

In 1957 Murayama called a compacted sand pile constructed in a ground as the 'Sand Compaction Pile'. Since then the method has been frequently called as the 'Sand Compaction Pile method', a term which would cover similar techniques developed later. The hammering compaction technique required large energy to compact sand piles in a ground, and the heavy impact caused considerable disturbance of the surrounding soil, which required additional sand [1]. Quality control in this technique was quite difficult especially when applied to clays, because the pore water pressure generated in a ground caused soils to intrude into the casing pipe [2]. Unfortunately,

Figure 6.1a.   SCP machine with hammering compaction technique in 1952 [3].

Figure 6.1b.   SCP machine with hammering compaction technique in 1955 [3].

the 'Compozer Method' with the hammering compaction technique was not adopted widely, because of frequent damage to machinery caused by the heavy impact of hammering, difficulties in quality control of constructing compacted sand piles, and severe noise during the hammering.

A new compaction technique was developed and adopted for the SCP method in 1959, in which a vibro-hammer was used to compact sand piles instead of the hammering device (vibrating compaction technique). In the technique, the vibro-hammer with a frequency of about 10 Hz was clamped to the upper end of a long casing pipe as shown in Figure 6.2. The sand fed into a ground was compacted by the vibrating excitation of the casing pipe in the vertical direction by means of the vibro-hammer. It was reported that the internal friction angle of sand subject to vibration decreased linearly with increasing logarithm of applied vibration acceleration [1]. This phenomenon meant that sand could be compacted sufficiently even by relatively small external load. In this technique, installation and compaction of sand in a ground could be carried out continuously, which made it possible to speed up the execution time. This technique had the other advantages, such as high quality and automatic control of constructing sand piles [2]. Due to these advantages, this technique has gradually

Figure 6.2.   SCP machine with vibrating compaction technique [3].

replaced the hammering compaction technique. When applied to sandy grounds, this technique was expected to function to densify loose sandy ground by the vibratory excitation during the phase of penetration and retrieval of the casing pipe. This technique was applied mainly to sandy grounds to reduce settlement and increase uniformity at that time, and to prevent liquefaction later.

Very soft clay deposits were often encountered in coastal areas, which caused insufficient bearing capacity and/or large settlement when constructing infrastructures of port and harbor facilities. These soft grounds used to be treated by a replacement method in those days. In the method, the soft soil was excavated partially or totally and good quality material (usually sand) was filled in the excavated portion to assure sufficient strength and/or compressibility. In 1961, the SCP method was also used to improve such soft clay grounds as a replacement method in which many compacted sand piles were constructed with a high replacement area ratio so that almost all clay was replaced by the compacted sand piles [4]. A SCP barge with a wooden scaffold as shown in Figure 6.3 was manufactured and was successfully applied to improve marine clay deposits at Saganoseki Port, Oita Prefecture. Following its successful application to marine clay deposits, the SCP method has been applied not only to sandy grounds but also to clay grounds in many construction projects including the Kansai International Airport, Haneda Airport and Trans-Tokyo Bay Highway projects.

Figure 6.3.   SCP barge at Saganoseki Port, Oita Prefecture [3].

Figure 6.4a.   SCP machines for on-land construction [3].

A large number of SCP applications in the 1970s, 1980s and 1990s encouraged several construction firms to develop their own new compaction techniques and devices for on-land construction and marine construction [5–6] (see Figures 6.4a to 6.4c). In on-land construction, SCP machines having a casing pipe leader have been used. In marine construction, on the other hand, SCP barges having multiple casing pipe leaders (usually three leaders) have been used for rapid construction of sand piles.

The Sand Compaction Pile method with a vibrating compaction technique has been widely adopted in many constructions. However, this technique can cause severe noise and vibration influences on the surrounding area. Many construction projects in urban areas required a new technique that could eliminate much of the noise and vibration. One of the solutions was the 'mini compozer method', which consisted of a casing pipe with relatively small diameter of 20 cm to 35 cm and a vibro-hammer with a high frequency of 20 Hz to 25 Hz (Figure 6.5 [3]). The diameter of sand piles constructed by this technique was about 35 cm to 50 cm. This technique was applied to a highway construction project in 1988 in which compacted sand piles were constructed with an interval of 1.1 m, which corresponded to the replacement area ratio, a_s, of 0.08. The vibration and noise during the sand piles installation were reduced 5 dB to 12 dB and 10 dB respectively, yet the improvement effect was almost the same as that by the vibrating compaction technique [7–8].

Figure 6.4b.   SCP barge for marine construction [3].

Figure 6.4c.   SCP barge for marine construction [3].

Figure 6.5.   SCP machine for 'mini compozer method' [3].

In 1995, a unique compaction technique was developed as shown in Figure 6.6, in which sand piles were compacted statically by rotating the casing pipe instead of vibratory excitation [9–10]. This technique can be categorized as a static compaction technique (non-vibrating compaction technique) and greatly reduces the severe noise and vibration. In this technique, the casing pipe is penetrated statically into a ground with rotation by means of a motor jack. The sand fed into a ground is statically compacted by rotating the casing pipe during the retrieval phase. It was found in many field measurements that the improvement effect by this technique was similar to that by the vibrating compaction technique [9–10]. Because no vibrating excitation is used in the machine, the noise and vibration during execution can be considerably reduced such that there are negligible adverse impacts on those living near the construction site (see Figures 6.7a and 6.7b [10]). This technique has high applicability to construction projects in urban areas, especially close to existing structures. Recently, this technique has been adopted for marine construction [11]. Similar static compaction techniques have been developed by several construction firms [12–15]. After developing the static compaction technique, research has focused on the possibility of applying by-product materials and subsoil to this technique as later described in Chapter 6.3.

For the static compaction technique (non-vibrating compaction technique), Ando *et al.* [16] performed laboratory tests to investigate the compaction mechanism due to static pressure. Ishihara *et al.* [17] attempted the compaction mechanism by an elastic stress path approach.

Beside the developments in SCP machines and techniques, execution and quality control systems have also been developed. In 1981, an automatically controlled SCP driving system was invented, which enabled vibration conditions to be controlled according to the properties of each soil. In 1988, an automated execution system was

Figure 6.6.   SCP machine for static compaction technique [3].

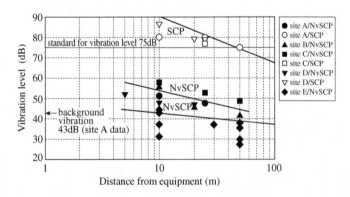

Figure 6.7a.   Vibration level during sand piles installation [10].

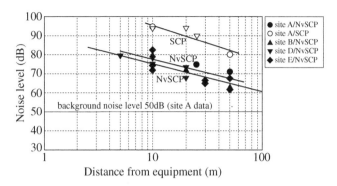

Figure 6.7b.   Noise level during sand piles installation [10].

developed for marine construction, in which almost all execution techniques including positioning, supplying and feeding sand, and compacting sand piles were able to be automatically controlled [18].

One important issue is accurately positioning a SCP machine to the prescribed position. In marine construction, the position of SCP barges used to be measured and controlled by transit observation from several fixed positions, and the SCP barge was manually positioned by controlling the anchor strength according to the transit observations. This technique was quite time-consuming. A semi-automatic system for positioning SCP barge was developed and put into practice in the 1980s. A fully-automatic system for positioning SCP barge was developed in 1996, which consisted of a computer-aided system coupled with the Global Positioning System (GPS). This technique is quite useful for marine construction, as it enables rapid positioning with high accuracy.

According to these developments of SCP machines, techniques and execution systems, the maximum improvement depth of the method had been increased to 70 m by 1993. The cumulative length of compacted sand piles also increased very rapidly in the 1970s, 1980s and 1990s, and reached 350 thousand km in 2001, as summarized in Figure 6.8. Nowadays, the SCP method is used to improve many kinds of ground including clay ground, sandy ground and fly ash ground for various improvement purposes.

6.3   MATERIAL DEVELOPMENT

As high strength and permeability of sand piles are expected for application to clay ground, sand materials suitable for the SCP method are well blended granular materials with low fines content and high particle strength. Recently, the limited availability of sand suitable for the method has necessitated research to find new materials. Many new materials including steel, copper and ferro-nickel slag, oyster shell and granulated coal fly ash have been used as SCP materials. For application to sandy grounds, on the other hand, permeability of sand piles is of less importance, which allows

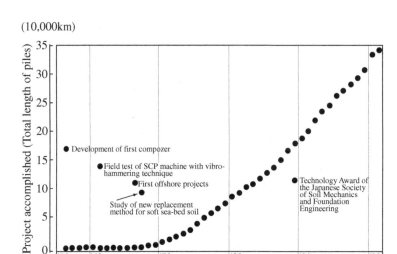

Figure 6.8.   Cumulative length of compacted sand piles.

Table 6.1.   New materials used for SCP method.

Material	Applied for	Reference
Gravel, crushed sand	Vibrating compaction technique	Okamoto et al. [20]
		Fukute et al. [21]
Crushed concrete	Vibrating compaction technique	Nakano et al. [22]
Steel slag	Vibrating compaction technique	Oikawa et al. [23]
		Matsuda et al. [24]
Copper slag	Vibrating compaction technique	Minami et al. [25]
		Kitazume et al. [26]
Ferro-nickel slag	Vibrating compaction technique	Kitazume and Yasuda [27]
Oyster shell	Vibrating compaction technique	Hashidate et al. [28–29]
Blast furnace slag	Static compaction technique	Yamamoto et al. [30]
Subsoil material	Static compaction technique	Matsuo et al. [31–32]
Mixture of slag, lime and aluminum	Static compaction technique	Ishii et al. [33]
Granular material	Static compaction technique	Fujiwara et al. [34]
Asphalt and concrete	Static compaction technique	Okado et al. [35]

subsoil materials or silty soils to be used as SCP materials. These new types of materials have been frequently adopted especially in the static compaction techniques. Some design manuals have been prepared for some types of steel slag [19]. Two case histories where copper slag or oyster shell was used as a SCP material are introduced in Chapter 5. Table 6.1 summarizes new materials used for the SCP method.

6.4   CURRENT TECHNIQUES

6.4.1   Classification of techniques

As introduced in the previous section, several types of SCP machine have been developed and adopted for on-land construction and marine construction for more than 50 years. However, some of them were replaced by new techniques, such as the hammering compaction technique that was the first development of the SCP method. Table 6.2 summarizes several execution types of SCP method that are currently available in Japan [36]. The techniques for executing the SCP method can be divided into three groups: hammering compaction technique, vibrating compaction technique and static compaction technique (non-vibrating compaction technique).

In the hammering compaction technique, sand fed out from a casing pipe is compacted by a similar manner to the pile driving technique, by dropping a heavy weight (ram) onto the sand in the casing pipe. As described in Chapter 6.2, the shock impact by hammering used to be used, but is now rarely used due to frequent damage to machinery, difficulties in quality control, and severe noise and vibration.

In the vibrating compaction technique, sand fed into a ground is compacted by vibratory excitation. The vibrating compaction technique has three variations according to the position and type of vibrating probe as summarized in Table 6.2: (a) vertical vibrating compaction technique, (b) vertical and horizontal vibrating compaction technique and (c) vibro-compaction-device technique. This technique has been frequently applied for both on-land construction and marine construction.

In the static compaction technique (non-vibrating compaction technique), on the other hand, sand introduced in a ground is compacted by static rotational and/or downward movements of a casing pipe or a probe. This technique has three variations according to the type of compaction: (a) rotation and wave compaction technique, (b) double casing pipes compaction technique and (c) rotary compaction-device technique. Since this technique has several advantages including quite low noise and vibration, it has been applied to on-land construction especially in urban areas, and recently to marine construction [11].

The three types of SCP technique are briefly explained in the following section.

6.4.2   Hammering compaction technique

The machine consists of a crawler crane, a casing pipe and a hammering probe. The casing pipe of 30 cm in diameter is penetrated into a ground. After reaching the prescribed depth, sand is fed into the ground by lifting the casing pipe to some amount and is compacted in a similar manner to the pile driving technique, by dropping a shock hammer (heavy weight ram) onto the sand in the casing pipe. This procedure is repeated to construct the sand pile to ground level. The hammering compaction technique requires large energy to compact sand piles in a ground, and the heavy impact causes considerable disturbance to the surrounding soil, which requires additional sand [1].

Table 6.2.   Various techniques of SCP method [36].

Type of compaction	Type of technique	Name of method
Hammering compaction technique	Shock hammering technique	Hammering SCP method
Vibrating compaction technique	Vertical vibrating compaction technique	Compozer method
	Vertical and horizontal vibrating compaction technique	SS-P method
	Vibro-compaction-device technique	KS-HARD method
Static compaction technique (non-vibrating compaction technique)	Rotation and wave compaction technique	SAVE compozer
	Double casing pipes compaction technique	Geo-KONG method, SDP method
	Rotary compaction-device technique	KS-EGG method

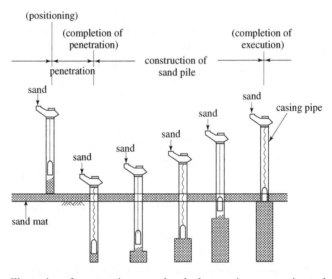

Figure 6.9.   Illustration of construction procedure by hammering compaction technique.

The improvement operation for the hammering compaction technique is shown in Figure 6.9 and is described as follows.

i) The casing pipe is located at the design position.
ii) The casing pipe is driven into a ground by the hammering probe.

iii) After reaching the prescribed depth, the casing pipe is retrieved about one meter to feed the sand filled in the casing pipe into the ground with the help of compressed air.

iv) The sand fed into the ground is compacted to expand the diameter by the hammering probe. The degree of compaction is controlled so that the diameter of the sand pile becomes the designed value.

v) After compacting the sand to the designed degree, the casing pipe is retrieved about one meter again and the sand is fed into the ground. The sand is compacted again by the hammering probe.

vi) These procedures are repeated until a compacted sand pile is constructed up to the ground level. During the procedure, sand or granular material is continuously supplied through the hopper by the lifting bucket. During the construction, the depth of the casing pipe and the position of the sand in the casing pipe are continuously measured for quality control and assurance.

6.4.3   Vibrating compaction technique

In the vibrating compaction technique, a casing pipe is driven into a ground by a vibro-hammer on the top of the casing pipe. While retrieving the casing pipe, sand fed into the ground is compacted by the vibratory excitation of the casing pipe. The vibrating compaction technique has three variations according to the position and type of vibrating probe as summarized in Table 6.2: (a) vertical vibrating compaction technique, (b) vertical and horizontal vibrating compaction technique and (c) vibro-compaction-device technique [36].

Vertical vibrating compaction technique
The vertical vibrating compaction technique has been most frequently used in SCP improvement. Typical SCP machines of the vertical vibrating compaction technique are shown in Figures 6.10a and 6.10b for on-land construction and marine construction respectively. The machine for on-land construction consists of a crawler crane with a leader and a casing pipe, a vibro-hammer and a compressor to supply high air pressure to a casing pipe. A crawler crane with a lifting capacity of 250 kN to 400 kN (25 tf to 40 tf) is often used. The casing pipe of 40 cm to 50 cm in diameter is suspended along the leader through the vibro-hammer to construct compacted sand piles whose diameter ranges from 50 cm to 70 cm. The vibro-hammer is suspended from the crane via the shock absorber. A lifting bucket is also suspended along the leader and goes up and down to supply granular material to the casing pipe through the hopper. Some types of casing pipe have a unique bottom plate as shown in Figures 6.11a and 6.11b, which functions to prevent the sand from flowing out of the casing pipe and to prevent the soil from being squeezed into the casing pipe during penetration of the casing pipe.

   As a kind of this technique, a 'mini compozer method' has been adopted especially in urban areas, which consists a casing pipe with small diameter of 20 cm to 35 cm and

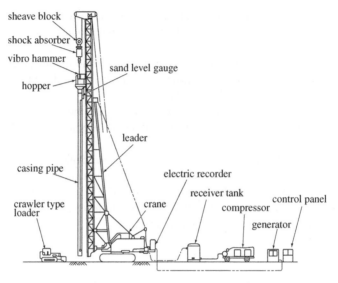

Figure 6.10a.   SCP machine of vertical vibrating compaction technique for on-land constructions.

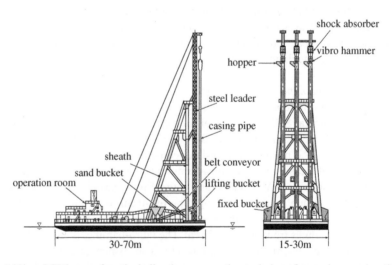

Figure 6.10b.   SCP barge of vertical vibrating compaction technique for marine constructions.

a vibro-hammer with high frequency of 20 Hz to 25 Hz. The diameter of sand piles constructed by this technique is about 35 cm to 50 cm [7–8].

The improvement operation for the vertical vibrating compaction technique is shown in Figure 6.12 and is described as follows.

Figure 6.11a.   Bottom plate of casing pipe.

Figure 6.11b.   Bottom plate of casing pipe.

i) The casing pipe is located at the design position.
ii) The casing pipe is driven into a ground by the help of vertical vibratory excitation by the vibro-hammer on the top of casing pipe. Compressed air is injected from the outlet nozzles installed on the side face of casing pipe in the case of penetrating relatively hard stratum. During the penetration, the casing pipe is filled with sand or granular material that is supplied through the hopper at the upper end of casing pipe by the lifting bucket.

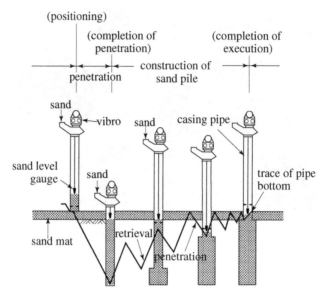

Figure 6.12.   Execution procedure for vertical vibrating compaction technique.

iii) After reaching the prescribed depth, the casing pipe is retrieved about one
meter to feed the sand filled in the casing pipe into the ground by the help of compressed air.

iv) The sand fed in the ground is compacted to expand diameter by the vibratory excitation of casing pipe in the vertical direction. The degree of compaction is controlled so that the diameter of sand pile becomes the designed value.

v) After compacting the sand to the designed degree, the casing pipe is retrieved one
meter again and fed the sand in the ground. The sand is compacted again by the
vibratory excitation.

vi) These procedures are repeated until a compacted sand pile is constructed up to the
ground level. During the procedure, sand or granular material is continuously supplied through the hopper by the lifting bucket. During the construction, the depth
of the casing pipe, the position of the sand in the casing pipe are continuously
measured for quality control and assurance.

Vertical and horizontal vibrating compaction technique
The machine for this technique is same as that for (1) the vertical vibrating compaction technique in principle, except a vibro-flot also being installed on the bottom
of casing pipe. The sand fed in a ground is compacted by the combination of vertical
and horizontal vibrations. The vertical vibration is provided by the vibro-hammer
installed on the top of casing pipe and the horizontal vibrations is provided by the
vibro-flot installed on the bottom of casing pipe.

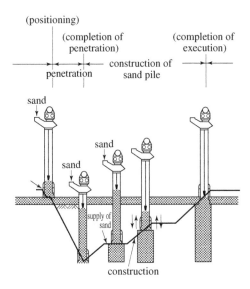

Figure 6.13.   Execution procedure for vertical and horizontal vibrating compaction technique.

The improvement operation for the technique is shown in Figure 6.13 and is explained as follows.

i)   The casing pipe is located at the design position.
ii) The casing pipe is driven into the ground by the help of vertical vibratory excitation by the vibro-hammer on the top of casing pipe. Compressed air is injected from the outlet nozzles installed on the side face of casing pipe in the case of penetrating relatively hard stratum. During the penetration, the casing pipe is filled with granular material that is supplied through the hopper at the upper end of the pipe by the lifting bucket.
iii) After reaching the prescribed depth, the casing pipe is gradually retrieved to feed the sand filled in the casing pipe into the ground by the help of compressed air.
iv) The sand fed in the ground is compacted by the combined functions of the vibro-flot on the bottom of casing pipe and the vibro hammer on the top of the casing pipe. The sand also expands in diameter. The degree of compaction is controlled so that the diameter of sand pile becomes the designed value.
v)   During retrieving the casing pipe continuously, feeding and compacting the sand are done.
vi) These procedures are repeated until the compacted sand pile is constructed up to the ground level. During the procedure, granular material is frequently supplied through the hopper by a lifting bucket. During the construction, the depth of the casing pipe, position of sand in the casing pipe is continuously measured for quality control and assurance.

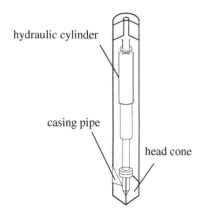

hydraulic cylinder

casing pipe

head cone

Figure 6.14.   Vibro-compaction-device.

Vibro-compaction-device technique
The machine for this technique is same as that for (1) the vertical vibrating compaction technique in principle, except a vibro-compaction-device being installed on the bottom of casing pipe. Figure 6.14 illustrates the vibro-compaction-device, which can push the head cone by hydraulic cylinder and vibration to compact sand in a ground. The casing pipe is driven into a ground by the vibro-hammer installed on the top of casing pipe. After then, the sand fed in a ground is compacted by the combined function of the vibro-compaction-device in the casing pipe and the vibro-hammer on the top of casing pipe. This technique was used to constructions, but is not used recently.

The improvement operation for the technique is shown in Figure 6.15 and is explained as follows.

i) The casing pipe is located at the design position.
ii) The casing pipe is driven into the ground by the help of the vertical vibratory exci-tation by the vibro-hammer on the top of casing pipe. Compressed air is injected from the outlet nozzles installed on the side face of casing pipe in the case of pen-etrating relatively hard stratum. During the penetration, the casing pipe is filled with granular material that is supplied through the hopper at the upper end of the pipe by the lifting bucket.
iii) After reaching the prescribed depth, the casing pipe is gradually retrieved to feed the sand filled in the casing pipe into the ground by the help of compressed air.
iv) The sand fed in the ground is compacted by the combined function of the vibro-compaction-device (installed in the casing pipe) and vibro-hammer. The sand also expands in diameter. The degree of compaction is controlled so that the diameter of sand pile becomes the designed value.
v) During retrieving the casing pipe continuously, feeding and compacting the sand are done.

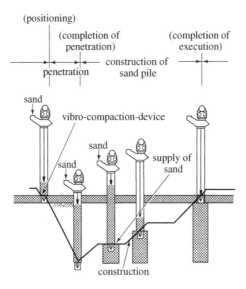

Figure 6.15.   Execution procedure for vibro-compaction-device technique.

vi) These procedures are repeated until the compacted sand pile is constructed up to the ground level. During the procedure, granular material is frequently supplied through the hopper by a lifting bucket. During the construction, the depth of the casing pipe, position of sand in the casing pipe is continuously measured for quality control and assurance.

6.4.4   Static compaction technique (non-vibrating compaction technique)

In the static compaction technique (non-vibrating compaction technique), a casing pipe is penetrated into a ground statically by means of a driving/lifting device. After reaching the prescribed depth, the casing pipe is retrieved and sand fed into the ground is compacted statically. In this category, three variations are listed: (a) rotation and wave compaction technique, (b) double casing pipes compaction technique and (c) rotary compaction-device technique.

Rotation and wave compaction technique
The static compaction assembly of this technique is schematically shown in Figure 6.16 [3]. The machine consists of a crawler crane with a leader and a casing pipe, a driving device, a rotary drive motor and a compressor to supply high air pressure to the casing pipe. The crawler crane with a lifting capacity of 250 kN to 400 kN (25 tf to 40 tf) is often used. The casing pipe with 40 cm in diameter is suspended along the leader through the driving device and the rotary drive motor to construct compacted sand piles whose diameter is 70 cm. The operation assembly consists of a driving device, an

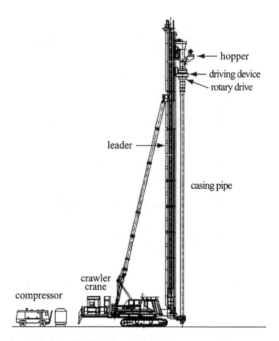

Figure 6.16.   Illustration of SAVE composer [3].

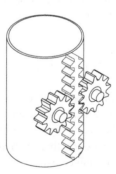

Figure 6.17a.   Pin lack & sprocket type driving/lifting device in static compaction technique
(non-vibrating compaction technique) [3].

electric rotary drive motor for rotating the casing pipe, and a driving/lifting device with
a rack-pinion type geared motor for the vertical movement of casing pipe. Figures
6.17a and 6.17b illustrate the driving/lifting device, in which two types of device, pin
lack & sprocket type and lack & pinion type, are manufactured. In both types, the gear
is driven by a hydraulic motor. The casing pipe is penetrated together with rotating
by means of the driving device and the rotary drive motor. A lifting bucket is also

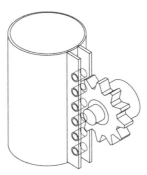

Figure 6.17b.   Lack & pinion type driving/lifting device in static compaction technique (non-vibrating compaction technique) [3].

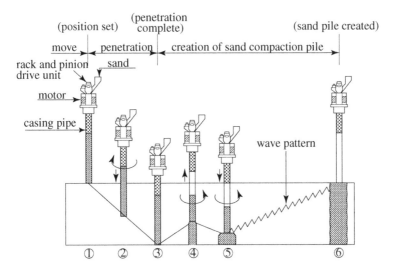

Figure 6.18.   Execution procedure for static compaction technique (non-vibrating compaction technique).

suspended along the leader and goes up and down to supply granular material to the casing pipe through the hopper. The casing pipe does not have a bottom plate.

The improvement operation for the technique is shown in Figure 6.18 and is explained as follows.

i) The casing pipe is located at the design position.
ii) The casing pipe is penetrated into the ground statically by means of the driving/lifting device with rotating the casing pipe by the rotary device. In this stage, the casing pipe is filled with sand or granular material. Compressed air is

injected from the outlet nozzles installed on the inner side face of casing pipe in the case of penetrating relatively hard stratum.

iii) After reaching the prescribed depth, the casing pipe is retrieved about a half meter and the sand or granular material filled in the casing pipe is fed into the ground by the help of compressed air.

iv) The sand fed in the ground is compacted by the downward and rotational movements of the casing pipe. The sands in the ground is expanded in diameter, which causes to make the surrounding soil dense.

v) After compacting the sand to the designed degree, the casing pipe is retrieved about a half meter again and fed the sand in the ground. The sand is compacted again by the downward and rotational movements of the pipe.

vi) These procedures are repeated until a compacted sand pile is constructed up to the ground level. During the procedure, sand or granular material is continuously supplied through the hopper by the lifting bucket. During the construction, the depth of the casing pipe, the position of the sand in the casing pipe are continuously measured for quality control and assurance.

The most benefit of the static compaction technique is that the machine can considerably reduce noise and vibration during the execution so that negligible adverse impacts cause to the surroundings (see Figure 5.46). According to the advantage, this techniques has high applicability to heavy populated urban area where many buildings are build close each other.

Double casing pipes compaction technique
Figure 6.19 shows an example of machine of double casing pipes compaction technique in the static compaction technique, which consists of a crawler crane with a leader and a casing pipe, an auger device [37]. A set of double casing pipes is installed in this technique, whose diameters are usually 50 cm and 55 cm for the inner and outer casing pipes respectively. The set of double casing pipes is penetrated into a ground by rotating the outer casing pipe. When a retrieval stage, the inner casing pipe goes up and push down back by means of a hydraulic cylinder to feed granular materials into a ground and compact it, as illustrated in Figure 6.20 [14].

The improvement operation for the technique is shown in Figure 6.21 and is explained as follows.

i) The casing pipe is located at the design position.

ii) The casing pipe is penetrated into the ground statically by rotating the outer casing pipe. In this stage, the casing pipe is filled with sand or granular material.

iii) After reaching the prescribed depth, the casing pipe is stopped to rotate.

iv) The sand fed in the ground is compacted by going up and pushing down the inner casing pipe by means of a hydraulic cylinder.

v) These procedures are repeated until a compacted sand pile is constructed up to the ground level. During the procedure, sand or granular material is continuously supplied through the hopper by the lifting bucket.

Figure 6.19.   Double casing pipes compaction technique [37].

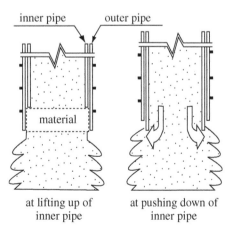

Figure 6.20.   Illustration of compacting sand by double casing pipes compaction technique [14].

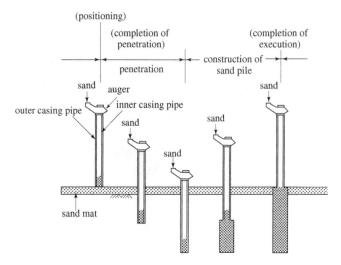

Figure 6.21.   Execution procedure for double casing pipes compaction technique.

Figure 6.22.   Compaction-device technique [38].

Rotary compaction-device technique
Figure 6.22 shows an example of machine of rotary compaction-device technique in the static compaction technique, which consists of a crawler crane with a leader and a casing pipe, a hydraulic rotary drive and a hydraulic winch [38]. A excavation and expanding diameter head is installed on the bottom of casing as shown in Figures 6.23a and

Figure 6.23a.   Photograph of excavation and expanding diameter head [38].

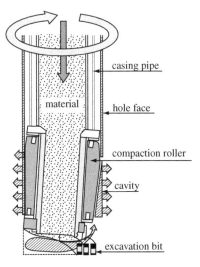

Figure 6.23b.   Illustration of excavation and expanding diameter head [13].

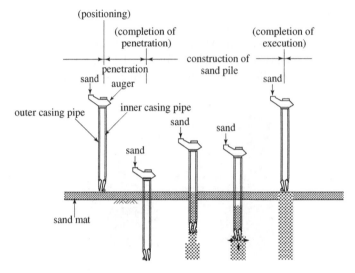

Figure 6.24.   Execution procedure for rotary compaction-device technique.

6.23b [13]. When the retrieval stage, a granular material fed into a ground is compacted by eccentric rotation of the head. The maximum improvement depth is −25 m.

The improvement operation for the technique is shown in Figure 6.24 and is explained as follows.

i) The casing pipe is located at the design position.
ii) The casing pipe is penetrated into the ground statically by rotating the casing pipe. In this stage, the casing pipe is filled with sand or granular material.
iii) After reaching the prescribed depth, the casing pipe is stopped to rotate.
iv) The sand fed in the ground is compacted by rotating the excavation and expanding diameter head.
v) These procedures are repeated until a compacted sand pile is constructed up to the ground level. During the procedure, sand or granular material is continuously supplied through the hopper by the lifting bucket.

REFERENCES

1 Murayama, S.: Considerations of Vibro-Compozer Method for application to cohesive ground. Journal of Construction Machinery, No.150, pp.10–15, 1962 (in Japanese).
2 Ogawa, M.: Development of Compozer Method. Journal of Construction Machinery, No.150, pp.16–19, 1962 (in Japanese).
3 The courtesy of Fudo Construction Co. Ltd.

4   Ogawa, M. and Koike, H.: Experiment on new replacement method of soft ground. Proc. of the 21st Annual Conference of the Japan Society of Civil Engineers, III, pp.120–1–2, 1966 (in Japanese).

5   Fudo Construction Co., Ltd.: Design manual for Compozer System. 1971 (in Japanese).

6   Nippon Kaiko Co., Ltd.: Design manual for S.S-P. method. Nippon Kaiko Co., Ltd., pp.27, (in Japanese).

7   Takahashi, Y., Koyamachi, N., Nitao, H. and Shigeno, M.: Application of sand compaction pile method with high frequency and small diameter. Journal of the Japanese Society of Soil Mechanics and Foundation Engineering, 'Tsuchi-to-Kiso', Vol.42, No.4, pp.45–48, 1994 (in Japanese).

8   Ando, Y., Tsuboi, H., Yamamoto, M., Harada, K. and Nozu, M.: Recent soil improvement methods for preventing liquefaction. Proc. of the 1st International Conference on Earthquake Geotechnical Engineering, pp.1–6, 1995.

9   Nozu, M., Ohbayashi, J. and Matsunaga, Y.: Application of the static sand compaction pile method to loose sandy soil. Proc. of the International Conference on Problematic Soils, pp. 751–755, 1998.

10 Tsuboi, H., Ando, Y., Harada, K., Ohbayashi, J. and Matsui, T.: Development and application of non-vibratory sand compaction pile method. Proc. of the 8th International Offshore and Polar Engineering Conference, pp. 615–620, 1998.

11 Otsuka, M. and Isoya, S.: save marine method. Journal of Construction Machine, No.8, pp. 58–61, 2003 (in Japanese).

12 Yamazaki, H., Morikawa, Y., Nakazato, T. and Tai, S.: Laboratory tests on application of subsoil to static compaction SCP method. Proc. of the 54th Annual Conference of the Japan Society of Civil Engineers, III, pp. 532–533, 1999 (in Japanese).

13 Kakehashi, T., Umeki, Y., Ookori, K. Soaji, Y., Makibe, Y. and Satou, M.: Technological introduction of low vibration and low noise type ground improvement construction method, – ecological gentle geo-improvement. Proc. of the 57th Annual Conference of the Japan Society of Civil Engineers, III, pp. 151–152, 2002 (in Japanese).

14 Kato, M., Tanaka, Y., Ichikawa, H. and Mishiro, N.: Geo-KONG Method. Journal of the Foundation Engineering & Equipment, 'Kisoko', Sogodoboku, No.12, pp. 38–41, 2003 (in Japanese).

15 Nishida, N., Watanabe, N., Hattori, M. and Shinoi, T.: Development of sand compaction pile method using inner-screw and high pressure intermittent air. (1) -abstract-. Proc. of the 38th Annual Conference of the Japanese Society of Soil Mechanics and Foundation Engineering, pp. 1057–1058, 2003 (in Japanese).

16 Ando, Y., Yamamoto, M., Harada, K. and Nozu, M.: Experimental study on densification of loose sandy soils by penetrating sand piles. Proc. of the 31st Annual Conference of the Japanese Society of Soil Mechanics and Foundation Engineering, pp. 73–74, 1996 (in Japanese).

17 Ishihara, K., Tsukamoto, Y., Satou, M., Harada, K., Yabe, H. and Amamiya, M.: Soil densification due to static sand pile penetration by hollow cylindrical torsional shear tests. Proc. of the 33rd Annual Conference of the Japanese Society of Soil Mechanics and Foundation Engineering, pp. 2157–2158, 1998 (in Japanese).

18 Shoji, A., Morisaki, K., Watanabe, K. and Goami, Y.: Automatic control system of SCP bessel. Journal of the Foundation Engineering & Equipment, 'Kisoko', Sogodoboku, No.4, pp. 88–92, 1988 (in Japanese).

19 Coastal Development Institute of Technology: Manual of utilization of steel slag. 146p., 2003 (in Japanese).

20 Okamoto, T., Ito, K., Nakajima, Y. and Okita, Y.: An executive example of gravel drain system in urban area. Proc. of the 21st Annual Conference of the Japanese Society of Soil Mechanics and Foundation Engineering, pp. 1867–1868, 1986 (in Japanese).

21 Fukute, T., Ohneda, H. and Tsuboi, H.: An experimental study on applicability of materials for compacted sand piles. Proc. of the 23rd Annual Conference of the Japanese Society of Soil Mechanics and Foundation Engineering, pp. 2139–2142, 1988 (in Japanese).

22 Nakano, K., Taniguchi, B. and Nakai, N.: The application of materials for non-vibratory sand compaction method (SAVE Compozer). Proc. of the 34th Annual Conference of the Japanese Society of Soil Mechanics and Foundation Engineering, pp. 1121–1122, 1999 (in Japanese).

23 Oikawa, K., Matsunaga, Y., Takahashi, K. and Hashimoto, T.: Applicability of steel-making slag and crushed concrete to SCP material. Proc. of the 32nd Annual Conference of the Japanese Society of Soil Mechanics and Foundation Engineering, pp. 2331–2332, 1997 (in Japanese).

24 Matsuda, H., Ando, Y., Kitayama, N. and Nakano, Y.: Study on effective utilization of granulated slag as a soil stabilizer. Proc. of the 33rd Annual Conference of the Japanese Society of Soil Mechanics and Foundation Engineering, pp. 2151–2152, 1998 (in Japanese).

25 Minami, K., Matsui, H., Naruse, E. and Kitazume, M.: Field test on sand compaction pile method with copper slag sand. Journal of Construction Management and Engineering, Japan Society of Civil Engineers, No.574/6-36, pp. 49–55, 1997 (in Japanese).

26 Kitazume, M., Shimoda, Y. and Miyajima, S.: Behavior of sand compaction piles constructed from copper slag sand. Proc. of the International Conference on Centrifuge Modeling, CENTRIFUGE 98, 1998.

27 Kitazume, M. and Yasuda, T.: Centrifuge model tests on horizontal resistance of pile in SCP improved ground with ferro-nickel slag. Proc. of the 35th Annual Conference of the Japanese Geotechnical Society, pp. 1441–1442, 2000 (in Japanese).

28 Hashidate, Y., Fukuda, T., Okumura, T. and Kobayashi, M.: Engineering properties of oyster shell-sand mixtures. Proc. of the 28th Annual Conference of the Japanese Society of Soil Mechanics and Foundation Engineering, pp. 869–872, 1993 (in Japanese).

29 Hashidate,Y., Fukuda, T., Okumura, T. and Kobayashi, M.: Engineering proper-
    ties of oyster shell-sand mixtures and their application to sand compaction piles.
    Proc. of the 29th Annual Conference of the Japanese Society of Soil Mechanics
    and Foundation Engineering, pp. 717–720, 1994 (in Japanese).
30 Yamamoto, Y., Takahashi, N., Kishishita, T., Hyodo, M., Miura, F., Saito, T. and
    Ikeda, R.: Liquefaction shaking table tests of SCP method using granulated coal
    ash. Journal of Earthquake Engineering, Japan Society of Civil Engineers,
    Vol.28, pp. 1–8, 2003 (in Japanese).
31 Matsuo, M., Kimura, M., Nishio, Y. and Ando, H.: Study on development of soil
    improvement method using construction waste soil. Journal of Geotechnical
    Engineering, Japan Society of Civil Engineers, No.547/3-36, pp. 199–210, 1996
    (in Japanese).
32 Matsuo, M., Kimura, M., Nishio, Y. and Ando, H.: Development of soil improve-
    ment method using construction waste soil. Journal of Construction Management
    and Engineering, Japan Society of Civil Engineers, No. 567/3-35, pp. 237–248,
    1997 (in Japanese).
33 Ishii, H., Horikoshi, K., Yamaguchi, J., Satoh, S. and Yamamoto, M.: Model tests
    for ground compaction behavior of new material for static compaction method.
    Proc. of the 35th Annual Conference of the Japanese Society of Soil Mechanics
    and Foundation Engineering, pp. 1427–1428, 2000 (in Japanese).
34 Fujiwara, T., Tanaka, H., Kadota, M., Ohiso, Y. and Kurata, T.: Field test of static
    compaction improved method by granulated excavated-soil. Proc. of the 35th
    Annual Conference of the Japanese Society of Soil Mechanics and Foundation
    Engineering, pp.1425–1426, 2000 (in Japanese).
35 Okado, M., Ogura, R., Wada, K. and Kobayashi, Y.: The application of recycled
    materials for static sand compaction pile method. Proc. of the 35th Annual
    Conference of the Japanese Society of Soil Mechanics and Foundation
    Engineering, pp. 1431–1432, 2000 (in Japanese).
36 Japanese Society of Soil Mechanics and Foundation Engineering: soil improve-
    ment methods – survey, design and execution. The Japanese Society of Soil
    Mechanics and Foundation Engineering, 1988 (in Japanese).
37 The courtesy of Konoike Construction Co., Ltd.
38 The courtesy of Japan Industrial Land Development Co., Ltd.

Chapter 7

Research and Development in Japan

7.1 INTRODUCTION

Since the SCP technique was first developed, many researches have been carried out by universities, academics and research firms, and construction and consulting firms in Japan. These efforts have revealed the strength and deformation characteristics of composite soils, bearing capacity, settlement, earth pressure and horizontal resistance of improved ground, liquefaction prevention and so on. Figure 7.1 summaries the research and development flow on the SCP method in Japan. Beside these researches, many types of execution machine have been developed and applied for many construction works as described in Chapter 6. In this chapter, research and development activities on the characteristics of improved ground in Japan are introduced in detail.

7.2 SHEAR STRENGTH OF COMPOSITE SOIL

The shear strength of composite soil, which consists of clay and sand piles, is influenced by many factors, such as the stress–strain characteristics of clay and sand piles, stress concentration effect, diameter and interval of sand pile, etc.

The first research on the shear strength of composite soil was carried out by Ibaragi [8], in which a series of direct shear tests was carried out on a composite soil specimen consisting of clay and a sand pile. In the tests, the soil type, replacement area ratio and vertical pressure were changed to investigate their effects on the shear strength characteristics. According to the tests, he proposed the shear strength formula given in Equation (7.1) that incorporated the stress concentration on the sand pile (Figure 7.2). Equation (7.1) means that the shear strengths of clay and sand piles are fully mobilized simultaneously. He also calculated the vertical stresses on soft soil and sand pile, σ_c and σ_s, by elastic theory.

$$
\begin{aligned}
S_0'' &= A_c \cdot \tau_{fc} + A_s \cdot \tau_{fs} \\
&= A_c \cdot \left(c_0 + \sigma_c \cdot \tan \phi_c \right) + A_s \cdot \sigma_s \cdot \tan \phi_s
\end{aligned}
\tag{7.1}
$$

	'50s	1960s	1970s	1980s	1990s	2000s
SCP machine	[1] [2]	[4] [12]				[154] [162][165] [163] [164]
new material			[35]		[105][110][119][129][111][130][112][131] [97][114][120][123][132][137][106][121][124][122]	[142] [143][146][144][145][151]
shear strength		[8] [13]			[82] [94] [107]	
bearing capacity stability		[5] [7] [10][14][20][21] [11] [15]	[23][25][29][30][36][31] [41]	[46][52][55][63][47][53][56][48][54][57] [50][51]	[78][80][83][88][91][98][113][124][132][84][89][94][99][115][85][90][95][108]	[147]
settlement, deformation		[5][6] [15]		[45][50][58] [66][67]	[99] [113] [133][138][139]	
stress concentration		[7]	[13][16][22][23][26][29][30][36][17][24] [39]	[49][51][61][59][62][60] [66]	[99] [113] [133][138][139]	
for clay — earth pressure			[38] [42]		[96] [125]	
horizontal resistance				[68][73]	[92][100][109][101][102][103][99]	[43][151][155][148][156][149]
execution		[5][6]	[32][33] [39][40] [43][44]	[64][69][74][81][86][70]	[116][117][123]	
design method		[10] [18][19]	[27][28] [37]	[59]	[84][93][85][87] [140][141]	[157] [168]
for sand — compaction mechanism	[3]				[126][134][135][118][136]	[152]
compaction effect		[9]	[34]	[65][71][75][72]	[127][136][104]	[152] [169][170]
design method		[9]	[27]	[76]	[128]	[150][153][158][161][159][160] [171]

Figure 7.1.   R/D flow on SCP method in Japan.

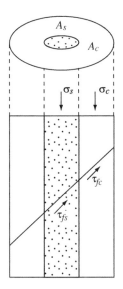

Figure 7.2.   Illustration of composite soil.

where:

A_c : cross sectional area of clay (m²)
A_s : cross sectional area of sand pile (m²)
c_0 : cohesion of clay (kN/m²)
S_0'' : shear strength of composite soil (kN)
σ_c : vertical stress on clay (kN/m²)
σ_s : vertical stress on sand pile (kN/m²)
τ_{fc} : shear strength mobilized in clay (kN/m²)
τ_{fs} : shear strength mobilized in sand piles (kN/m²)
ϕ_c : internal friction angle of clay
ϕ_s : internal friction angle of sand pile

Matsuo *et al.* [13] carried out a series of triaxial compression tests on composite soil and concluded that the shear strength of composite soil could be evaluated by the weighted sum of shear strengths of clay and sand piles mobilized at the same axial strain level as shown in Equation (7.2). They also pointed out that the shear strength mobilized in the sand pile was equal to or less than the shear strength of sand alone. They introduced a parameter, α, to incorporate the mobilization degree of shear strength of the sand pile. They concluded that Equation (7.2) provided a good estimation of the test results with $\alpha = 0.5$ for unconsolidated-undrained tests and with $\alpha = 1$ for isotropic consolidated-undrained tests. This means that the maximum shear strengths of clay and sand piles do not mobilize simultaneously. The stress concentration effect was not incorporated in their proposal.

$$(\sigma_1 - \sigma_3)_{s+c} = \frac{\alpha \cdot (\sigma_1 - \sigma_3)_s \cdot A_s + (\sigma_1 - \sigma_3)_c \cdot A_c}{A} \qquad (7.2)$$

where:

A	: cross sectional area of composite soil (m^2)
A_c	: cross sectional area of clay (m^2)
A_s	: cross sectional area of sand pile (m^2)
α	: a parameter with respect to mobilization degree of shear strength of sand pile
$(\sigma_1 - \sigma_3)_{s+c}$	: deviator stress mobilized in composite soil (kN/m^2)
$(\sigma_1 - \sigma_3)_s$	: deviator stress mobilized in sand pile (kN/m^2)
$(\sigma_1 - \sigma_3)_c$	: deviator stress mobilized in clay (kN/m^2)

Yagi *et al.* [51] carried out a series of simple shear tests and direct shear tests to investigate the effects of replacement area ratio and number of sand piles on the shear strength characteristics of composite soil. They found that the shear strength of composite soil was influenced by not only the replacement area ratio but also the number of sand piles. They concluded the reason for the phenomenon was the non-uniform shear strength distribution within sand piles.

Yoshikuni *et al.* [50] carried out undrained triaxial compression tests on composite soils and found that the drainage condition within the specimen caused the specimen size effect of shear strength.

Enoki *et al.* [82] performed laboratory tests on composite soils and found that the sand piles and clay did not fail simultaneously but failed individually. They proposed that the shear strength of composite soil should be evaluated as an anisotropic c-ϕ material based on the shear strength criteria of clay and sand piles individually.

Asaoka *et al.* [94,107] performed triaxial compression tests on composite soils and found that the shear strength of composite soil was dependent upon the drainage condition of the sand pile. They found that the shear strength of sand pile was much higher in an undrained condition than in a drained condition. They proposed that the shear strength of composite soil was the sum of the drained shear strength of sand pile and undrained shear strength of clay, and emphasized the importance of considering the drainage condition of sand piles when evaluating the bearing capacity of SCP improved ground.

7.3   BEARING CAPACITY AND STABILITY OF IMPROVED GROUND

7.3.1   Bearing capacity of isolated sand pile

In researches on the bearing capacity and stability of SCP improved ground, the stress distribution in soft soil and sand pipes is a key issue for evaluation. In one of the first researches, Murayama (1957) considered that the strength increase of soft ground was caused by the drainage function of sand piles and by installing sand piles. In 1962 he

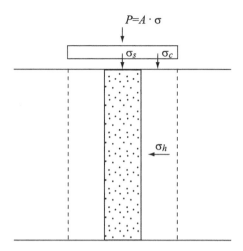

Figure 7.3.   Illustration of improved ground.

proposed a bearing capacity formula for improved ground by incorporating the stress concentration effect on sand piles.

Murayama [5] studied the bearing capacity and settlement behavior of SCP improved ground, and proposed Equations (7.3) to (7.5) for the bearing capacity calculation of an isolated sand pile by assuming an ultimate active state of sand pile and an ultimate passive state of surrounding soil (Figure 7.3). The principle of his calculation was based on the assumption that sand piles and surrounding clay were subjected to equal vertical deformation. In Equation (7.4), an upper yield stress of clay, σ_u, was introduced. He assumed σ_u as being equal to $0.7q_u$ by considering that no infinite creep deformation took place in the surrounding clay, where q_u was the unconfined compressive strength of clay. A similar equation was proposed by Hughes and Withers [172] for the bearing capacity of stone column improved ground.

(1) The vertical stress:

$$
\begin{aligned}
P &= A \cdot \sigma \\
&= (A_s + A_c) \cdot \sigma \\
&= A_s \cdot \sigma_s + A_c \cdot \sigma_c \\
&= \sigma_c (A_s \cdot n + A_c)
\end{aligned}
\tag{7.3}
$$

(2) the horizontal stress should be satisfied the following equation:

$$
\sigma_h \geq \frac{1 - \sin \phi_s}{1 + \sin \phi_s} \cdot \sigma_s
\tag{7.4a}
$$

$$
\sigma_h \leq \sigma_c + \sigma_u
\tag{7.4b}
$$

(3) the stress concentration ratio can be expressed as:

$$\frac{\sigma_s}{\sigma_c} = n$$

$$= \frac{1 + \sin \phi_s}{1 - \sin \phi_s} \cdot \left(1 + \frac{\sigma_u}{\sigma_c}\right) \tag{7.5}$$

where:

A : cross sectional area of composite ground (m^2)
A_c : cross sectional area of clay (m^2)
A_s : cross sectional area of sand pile (m^2)
σ : average vertical stress on composite ground (kN/m^2)
n : stress concentration ratio
P : bearing capacity (kN)
σ_c : vertical stress on clay (kN/m^2)
σ_s : vertical stress on sand pile (kN/m^2)
σ_h : horizontal confining stress on cylindrical surface of sand pile (kN/m^2)
σ_u : upper yield stress of clay (kN/m^2)
ϕ_s : internal friction angle of sand pile

Aboshi *et al.* [23] objected to Murayama's assumption regarding the stress condition of surrounding clay. They pointed out that Murayama's assumption meant that larger horizontal deformation should take place than vertical deformation, even if the composite ground was subjected to vertical loading.

Some field loading tests were performed by Niimi *et al.* [10] and model tests were performed by Hughes and Withers [172] to investigate the bearing capacity of an isolated sand pile experimentally.

7.3.2   Bearing capacity and stability of improved ground with multi sand piles

The first laboratory model tests were carried out in 1969 to investigate the bearing capacity of improved ground, where shear failure was found in some sand piles [20].

The bearing capacity of SCP improved ground with multi sand piles has been studied numerically by two approaches: by Terzaghi's theory and by the slip circle analysis.

In Terzaghi's theoretical approach for practical design, the bearing capacities of sand piles and clay were calculated individually as Equation (7.7) and the bearing capacity of SCP improved ground was calculated by weight averaging these bearing capacities as Equation (7.7a) while incorporating the replacement area ratio, *as*.

$$P = q_a \cdot A \tag{7.6}$$

$$q_a = as \cdot q_{as} + (1 - as) \cdot q_{ac} \tag{7.7a}$$

$$q_{as} = \frac{1}{Fs} B \cdot \gamma_s \cdot \beta \cdot N_\gamma + \gamma_s \cdot D \cdot N_q \qquad (7.7b)$$

$$q_{ac} = \frac{1}{Fs} c \cdot N_c \qquad (7.7c)$$

where:
- as : replacement area ratio
- A : cross sectional area of foundation (m²)
- B : width of foundation (m)
- c : undrained shear strength of clay between sand piles (kN/m²)
- D : embedded depth of foundation (m)
- Fs : safety factor (2.5 for static condition and 2.0 for dynamic condition)
- N_c : bearing capacity factor for cohesion
- N_q : bearing capacity factor for overburden pressure
- N_γ : bearing capacity factor for internal friction
- P : bearing capacity (kN)
- q_a : bearing capacity of improved ground (kN/m²)
- q_{as} : bearing capacity of uniform sandy ground having same property as sand piles (kN/m²)
- q_{ac} : bearing capacity of clay ground (kN/m²)
- β : shape factor of foundation
- γ_s : unit weight of sandy ground (kN/m³)

For the slip circle analysis approach, Murayama [5] studied the bearing capacity and settlement behavior of composite ground. He proposed the shear strength formula for SCP improved ground against sliding failure as Equation (7.8) (see Figure 7.4).

$$N = V \cdot \cos \qquad (7.8a)$$

$$T = N \cdot \tan \phi_e + c_e \cdot (A - A_s) \cdot \sec \beta \qquad (7.8b)$$

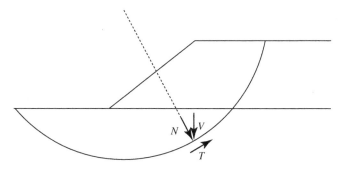

Figure 7.4.   Illustration of shear strength against slip failure.

where:

A : cross sectional area of composite ground (m^2)

A_s : cross sectional area of sand piles (m^2)

c_e : cohesion of clay mobilized on slip surface when sand piles and clay are sheared simultaneously

N : force perpendicular to slip surface (kN)

T : shear force mobilized in composite ground (kN)

V : vertical force on slip surface (kN)

β : inclination angle of slip surface

ϕ_e : internal friction angle of sand pile mobilized on slip surface

Matsuo et al. [20] and Matsuo [21] proposed a slip circle analysis for evaluating the bearing capacity of SCP improved ground. In their proposal, two safety factors were calculated individually by the slip circles passing through sand piles and clay ground, and through the clay portion alone, and the final safety factor was obtained by iterative calculations changing the position and diameter of slip circles until the two safety factors coincided. They calculated the bearing capacity by this slip circle analysis coupled with the stress distribution by Kogler's theory, and found that their model test results provided good estimations.

Beside the slip circle analyses, Nakanodou [57], Ishizaki [95,108] and Asaoka et al. [88,94] performed elasto-plastic FEM analyses. Ishizaki [95,108] analyzed the bearing capacity of group column type improved ground by two-dimensional FEM and by quasi-three-dimensional FEM analyses with 'Multi-link element' to demonstrate the applicability of his calculation model to bearing capacity. Asaoka et al. [88,94] performed FEM analyses to investigate the drainage condition of sand piles and emphasized the importance of the drainage condition of sand piles when evaluating the bearing capacity of SCP improved ground.

Regarding experimental approaches, Saitoh et al. [52] and Kimura et al. [63] carried out a series of centrifuge model tests to investigate the vertical and inclined bearing capacity of improved ground, and they found a sort of failure envelope in the vertical and horizontal loads plane [52]. Mikasa et al. [53] also carried out centrifuge model tests on the slope stability of improved ground.

In 1985, a full-scale loading test was conducted at Maizuru Port, Kyoto Prefecture. In the test, a SCP improved ground was subjected to vertical load until it failed. The test program and results are introduced in detail in Chapter 5.3. The test revealed that the bearing capacity of improved ground could be evaluated accurately by the slip circle analysis of the Fellenius method coupled with the shear strength formula (Equation (2.16a)) and Boussinesq's stress distribution method [99].

Terashi et al. [84–85,93] carried out a series of centrifuge model tests on the bearing capacity of improved ground subjected to vertical and horizontal loads. They observed the failure mode of sand piles in detail, and found that the failure envelope had a spindle shape in the vertical and horizontal loading planes similar to that of sandy ground. They also revealed that the Fellenius method coupled with the shear strength formula (Equation (2.16a)) could evaluate the model test results accurately.

The stability of sea revetments on SCP improved ground was investigated by centrifuge model tests [91,115,124,147]. Kitazume *et al.* [115] found that the sand piles in the backfill loading did not show shear failure but bending deformation.

7.3.3   Design formula for stability of clay ground

In addition to many physical and numerical model researches, the shear strength formula and slip circle analysis have been investigated for establishing practical design methods. Murayama [5] studied the bearing capacity and settlement behavior of SCP improved ground, and proposed a shear strength formula for improved ground against sliding failure as Equation (7.8).

Taguchi [11] provided a method of calculating the resistance moment of SCP improved ground, in which the shear strength increase of clay and the stress concentration effect on the shear strength of sand piles were taken into account. In his calculation, the effect of stress concentration was adopted for the external load. He proposed a specific design method of improved ground based on a consideration of the stress distribution of external loads.

Matsuo *et al.* [20] and Matsuo [21] proposed a slip circle analysis for evaluating the bearing capacity of SCP improved ground in which the three-dimensional installation pattern of sand piles was incorporated. In their proposal, the safety factors were calculated by slip circle analyses passing through sand piles and clay ground, and through the clay portion alone, and the final safety factor was obtained by iterative calculations until the two safety factors coincided. They calculated the stability by this slip circle analysis coupled with the stress distribution by Kogler's theory and found that their model test results provided good estimations.

Although there were some reports that the shear strengths in sand piles and clay did not mobilize simultaneously [13], the assumption of full mobilization of shear strength both in sand piles and clay has usually been adopted for establishing practical design methods. Nakayama [18] proposed a shear strength formula for SCP improved ground as Equation (7.9) according to the above-mentioned assumption (Figure 7.5). In the equation, the shear strength increase due to consolidation by external stress was taken into account. He provided a chart for calculating the slope stability problem of SCP improved ground for practical design use.

$$
\tau_{sc} = as \cdot \left(p\frac{\sigma_s}{\sigma} + \gamma_s \cdot z \right) \cdot \tan\phi_s \, \cos^2\theta
$$
$$
+ (1 - as) \cdot \left(c_0 + p\frac{\sigma_c}{\sigma} \cdot U \cdot \tan\phi \right) \qquad (7.9)
$$

where:
  as : replacement area ratio
  c_0 : cohesion of clay (kN/m^2)
  p : average vertical stress (kN/m^2)
  U : degree of consolidation

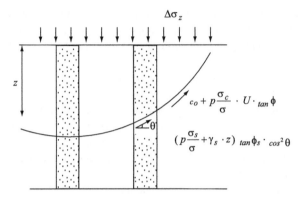

Figure 7.5.   Illustration of shear strength of improved ground.

z : depth (m)
γ_s : unit weight of sand pile (kN/m³)
θ : inclination of slip surface
σ : vertical stress on composite ground (kN/m²)
σ_c : vertical stress on clay (kN/m²)
σ_s : vertical stress on sand pile (kN/m²)
τ_{sc} : shear strength of improved ground (kN/m²)
ϕ_c : internal friction angle of clay ground
ϕ_s : internal friction angle of sand pile

Kuno and Nakayama [19] followed Nakayama's research [18] and carried out parametric studies of slip circle analysis on the slope stability problem. They included the effect of soil parameters such as clay strength, internal friction angle of sand piles, stress concentration ratio and so on, in the safety factor of the slip circle calculation. They also provided the optimum location of SCP improved ground based on slip circle analyses. A similar calculation was later conducted by Nakanodo et al. [57] by slip circle analyses and elasto-plastic FEM analyses. They found that the most effective location of improved ground was slightly different from that of the other two analyses.

Nakayama et al. [28] calculated the safety factors for 15 case histories by the slip circle analysis combined with the proposed shear strength formula as Equation (7.10). Equation (7.10) did not take into account the shear strength increase of clay due to external loading. They concluded that the slip circle analysis with the proposed equation was able to evaluate the stability of SCP ground accurately.

$$\tau_{sc} = as \cdot \left(p \frac{n}{(n-1) \cdot as + 1} + \gamma_s \cdot z \right) \cdot \tan \phi_s \, \cos^2 \theta + (1 - as) \cdot c$$

$$(7.10)$$

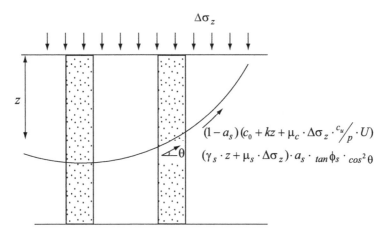

Figure 7.6.   Illustration of shear strength of improved ground for Formula (1).

where:

    as : replacement area ratio
    c : cohesion of clay (kN/m²)
    n : stress concentration ratio
    p : average vertical stress (kN/m²)
    z : depth (m)
    γ_s : unit weight of sand pile (kN/m³)
    θ : inclination of slip surface
    τ_{sc} : shear strength of improved ground (kN/m²)
    ϕ_s : internal friction angle of sand pile

These investigations were characterized by the shear strength formula proposed by Murayama and Ibaragi, and the stress distribution was calculated according to Boussinesq's elastic solution (see Chapter 2.4.1).

Sogabe [43] summarized the shear strength formulas as shown in Equations (7.11a) to (7.11e) and the statistics of adopting each formula in the practical design for constructing port and harbor facilities. Figure 7.6 illustrates the shear strength of improved ground for Formula (1). He also compared the safety factors calculated by each shear strength formula. Similar studies were later carried out by Kanda and Terashi [87].

Formula (1):

$$\tau = (1 - as) \cdot (c_0 + kz + \mu_c \cdot \Delta\sigma_z \cdot c_u / p \cdot U)$$
$$+ (\gamma_s \cdot z + \mu_s \cdot \Delta\sigma_z) \cdot as \cdot \tan\phi_s \cdot \cos^2\theta \qquad (7.11a)$$

Formula (2):

$$\tau = (1 - as) \cdot (c_0 + kz)$$
$$+ (\gamma_m \cdot z + \Delta\sigma_z) \cdot \mu_s \cdot as \cdot \tan\phi_s \cdot \cos^2\theta \qquad (7.11b)$$

Formula (3):

$$\tau = (\gamma_m \cdot z + \Delta\sigma_z) \cdot \tan\phi \cdot \cos^2\theta \qquad (7.11c)$$

Formula (4):

$$\tau = (\gamma_m \cdot z + \Delta\sigma_z) \cdot \tan\phi_m \cdot \cos^2\theta \qquad (7.11d)$$

Formula (5):

$$\tau_s = (\gamma_s \cdot z + \mu_s \cdot \Delta\sigma_z) \cdot \tan\phi_s \cdot \cos^2\theta$$
$$\tau_c = c_0 + kz \qquad (7.11e)$$

where:

  as : replacement area ratio
  c_0 : shear strength of clay at ground surface (kN/m²)
  c_u/p : shear strength increment ratio
  k : increment ratio of shear strength of clay with depth (kN/m³)
  n : stress concentration ratio

$$n = \frac{\sigma_s}{\sigma_c}$$

  U : average degree of consolidation
  z : depth (m)
  $\Delta\sigma_z$: increment of vertical load intensity (kN/m²)
  γ_s : unit weight of sand pile (kN/m³)
  γ_m : average unit weight of improved ground (kN/m³)

$$\gamma_m = \gamma_s \cdot as + \gamma_c \cdot (1 - as)$$

  θ : inclination of slip circle
  μ_c : stress concentration coefficient of clay ground for external load

$$\mu_c = \frac{\sigma_c}{\sigma} = \frac{1}{1 + (n - 1) \cdot as}$$

  μ_s : stress concentration coefficient of sand pile for external load

$$\mu_s = \frac{\sigma_s}{\sigma} = \frac{n}{1 + (n - 1) \cdot as}$$

σ_c : vertical stress on clay ground (kN/m²)
σ_s : vertical stress on sand pile (kN/m²)
τ : shear strength of improved ground (kN/m²)
ϕ : internal friction angle of sand
ϕ_m : internal friction angle of sand pile

$$\phi_m = \tan^{-1}(\mu_s \cdot as \cdot \tan \phi_s)$$

ϕ_s : internal friction angle of sand pile

In 1985, a full-scale loading test was conducted at Maizuru Port, Kyoto Prefecture (see Chapter 5.3). The test revealed that the bearing capacity of improved ground could be evaluated accurately by the slip circle analysis of the Fellenius method coupled with the shear strength formula (Equation (7.11a)) and the stress distribution method.

Kanda and Terashi [87] studied the case histories on the selection of design formulas and carried out parametric studies on the effect of design parameters on safety factors. A detailed description is given in Chapter 8.2. Terashi and Kitazume [85] and Terashi et al. [84] performed a series of centrifuge model tests on the bearing capacity of improved ground and found that the slip circle analysis with the shear strength formula (Equation (7.11a)) provided accurate estimations of the test results. Shinsha et al. [90] also confirmed in their centrifuge model tests that the slip circle calculation with the shear strength formula (Equation (7.11a)) was applicable to the stability of embankment on SCP improved ground. According to these confirmations, the slip circle analysis combined with the shear strength formula (1) (Equation (7.11a)) has been standardized in the practical design for port and harbor facilities [140,173].

Selection of design parameters was another important issue for accurate evaluation of bearing capacity and stability of SCP improved ground. Yabushita et al. [37] provided a sort of standard value of stress concentration ratio and internal friction angle of sand piles for several replacement area ratios according to the previous case histories. Ichimoto et al. [44] studied the selection of design parameters according to the case histories of practical design and field measurements. The effect of design parameters (soil parameters) incorporated in shear strength formula on the safety factor has been investigated several times [19,48,87]. Based on these research results, these standard values were summarized as shown in Tables 2.1 and 2.2.

7.4   SETTLEMENT OF IMPROVED GROUND

At the beginning of SCP development, the consolidation phenomenon of SCP improved ground was assumed to be the same as that of ground improved by the vertical drain method, and the consolidation process of the improved ground could be calculated by Barron's equation [5].

Matsuo *et al.* [13] showed that the consolidation process of SCP improved ground could be adequately evaluated in practice by Barron's theory provided the coefficient of consolidation of clay was properly estimated.

Ogawa and Ichimoto [6] accumulated field data on ground settlement and found that ground settlement was reduced by the stress concentration on sand piles and the consolidation effect of sand piles installation.

Matsuo *et al.* [13,15] carried out a series of triaxial compression tests and laboratory model tests on composite soils and found that the coefficient of consolidation changed during the consolidation process and the shape of the consolidation curve was different from that of Barron's theory. They considered that the phenomenon was caused by the dilatancy effect of sand piles. Takada and Nogawa [45] also found a similar phenomenon in their model tests, that the shape of the consolidation settlement curve of SCP improved ground had a moderate curvature in a semi-logarithm diagram as shown in Figure 7.7, which was different from Barron's theoretical curve [45].

Yoshikuni [41] derived a theoretical formula for the consolidation phenomenon by elastic theory, which incorporated the stress concentration effect of sand piles. Figure 7.8 shows a typical example of the theoretical consolidation phenomenon in

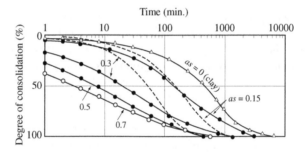

Figure 7.7.   Effect of replacement area ratio on time settlement curve of improved ground [45].

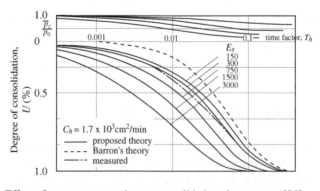

Figure 7.8.   Effect of stress concentration on consolidation phenomenon [25].

which the shear modulus of the sand pile is 13 times larger than that of the surrounding clay [25]. The figure indicates that the consolidation proceeds faster in the improved ground with increase of stress concentration. It also shows that the time settlement curves of SCP improved grounds have more moderate inclination than that by Barron's equation. He emphasized the importance of stress concentration effect on consolidation. The accuracy of their theoretical formula was confirmed by field settlement measurements of oil tanks [31].

As described above, theoretical calculation and some laboratory tests have shown that the consolidation phenomenon proceeds faster in SCP improved grounds than as predicted by Barron's theory. However, the accumulated field data have revealed the opposite phenomenon that the consolidation speed of SCP improved ground is slower than that estimated by Barron's equation, which is considered to be due to the soil disturbance effect. The relationship between the consolidation process and the replacement area ratio was given as shown in Figures 2.15 and 2.16 in 1981 and 1990.

Enoki *et al.* [66] carried out laboratory tests on composite soil and found that sand piles and clay showed three-dimensional deformation instead of one-dimensional deformation assumed in the practical design method.

Jung *et al.* [133,138] carried out a series of centrifuge model tests and laboratory model tests on the consolidation behavior of floating type improved ground, in which the effect of penetration depth of sand piles on the stress concentration ratio and consolidation phenomenon were investigated.

Oda and Matsui [139] conducted elasto-visco-plastic FEM analyses on one-dimensional consolidation of composite ground and compared the results with their model test results. They found that the stress condition in the ground was influenced by the dilatancy effect of sand piles and the compressible characteristics of clay. They showed that Barron's theory could predict the consolidation time accurately provided the permeability and/or compressibility of clay was suitably changed during the consolidation process.

In the practical design [140,173], the magnitude of consolidation settlement of improved ground is calculated by multiplying the consolidation settlement of original (unimproved) ground and a settlement reduction factor as shown in Figure 2.14, and the degree of consolidation is calculated by Barron's theory together with modifying the coefficient of consolidation (see Chapter 2.5).

7.5   STRESS CONCENTRATION RATIO

Many researches have been performed to investigate the stress concentration theoretically and experimentally.

From theoretical approaches, Matsuo *et al.* [13] provided the stress concentration ratio of 2.5 to 6 by a simple theoretical approach with an assumption that the horizontal stress acting on the boundary of sand piles and clay was equal to that of the outer surface of clay.

Yamaguchi and Murakami [39] performed a theoretical study on the stress concentration ratio, in which they considered three cases of stress condition of clay: elastic, elastic with partial plastic and fully plastic conditions. They derived theoretical values of stress concentration ratio for three cases, and compared them with the experiments by Mogami et al. [16–17].

Murayama et al. [29–30,36] carried out numerical calculations on SCP improved ground by the stress and strain relation of sand piles and clay. They compared their analyses with the experiment results and concluded that their analysis results coincided well.

Fukumoto and Yamagata [61] derived a theoretical equation for the stress concentration ratio, which incorporated the cohesive stresses mobilized on the interface of sand piles and clay. In their calculation, the stress concentration ratio in an undrained condition was derived based on an assumption of elastic behavior of sand piles and clay.

Enoki et al. [66] derived another theoretical equation for the stress concentration ratio by assuming a simple strain condition in a composite soil.

From experimental approaches, a first loading test was carried out by Motouchi et al. [7] to investigate the stress concentration ratio. Mogami et al. [16–17] performed laboratory model tests to investigate the effect of sand pile length and diameter, number of piles, magnitude of loading pressure and shear strength of clay on the stress concentration ratio. They found that the stress concentration ratio decreased with increasing loading pressure.

Matsuo et al. [22,26] carried out laboratory model tests on the stress distribution within improved ground. They found that the stress concentration ratio increased with increasing loading pressure, which was opposite to the phenomenon found by Mogami et al. [16–17]. They also found that the stress concentration ratio rapidly decreased after failure of the sand pile and decreased with increasing depth. They measured the horizontal stress in clay and a sand pile, and found that the stress ratio, $K = \sigma_h / \sigma_v$, decreased in the sand pile but increased in the clay during the loading.

Aboshi et al. [23] carried out a series of triaxial compression tests on composite soil and found that the stress concentration ratio increased to 4 to 7 as consolidation proceeded. These findings were also confirmed in the field measurements by Aboshi et al. [24].

Takechi et al. [49] and Yagi et al. [58] performed laboratory model tests on the stress distribution in improved ground, and found that the stress concentration ratio was influenced by the initial density of the sand pile and increased as consolidation proceeded. They also found that the stress distribution was not uniform within sand piles.

Enoki et al. [66] carried out a series of laboratory tests on composite soils and found that the stress concentration ratio for external load increment decreased with increasing loading pressure until the sand pile failed.

Jung et al. [133] carried out a series of centrifuge model tests on the effect of the penetration depth of sand pile on the stress concentration ratio and consolidation phenomenon. They found that the stress concentration ratio decreased with increasing penetration depth and changed with the degree of consolidation. They also found that the stress concentration ratio was higher at the ground surface than in the middle of improved ground.

Hirao [60] investigated the stress concentration ratio by field measurements at Maizuru Port and found that the magnitude of stress concentration ratio was dependent on the type of measurement.

Many researches have shown that the stress concentration ratio is greatly influenced by soil properties (strength and dilatancy characteristics) and initial stress condition of clay and sand piles, magnitude of external load, overburden pressure, etc. Many measured data have been accumulated in the laboratory, in model tests and in the field, but which shows opposite and different phenomena among each other. Among them, the effect of external pressure on the ratio was summarized by Suematsu [59], in which quite different phenomena in the relationship between the magnitude of external pressure and the ratio were shown.

Beside the effect of soil characteristics, the method of evaluating the stress concentration ratio is important. The ratio has often been evaluated by one of three methods: calculation of direct measurements of stress on sand piles and clay, back calculation by settlement reduction, and back calculation by shear strength increment of clay ground. Okada [99] showed different values of stress concentration ratio by the three methods in the field loading test at Maizuru Port, in which the ratios were 1.6, 3.2 and 2.53 for above three methods respectively.

In conclusion, the stress concentration ratio is a key parameter for the design of bearing capacity, stability and settlement and has been investigated by many researchers. However, further research is required to clarify it, both experimentally and theoretically.

7.6   EARTH PRESSURE OF IMPROVED GROUND

The number of researches on earth pressure of SCP improved ground is quite limited. Among them, Yabushita et al. [38] carried out laboratory model tests on the passive earth pressure of SCP improved ground. Terashi et al. [96] and Kitazume et al. [125] performed a series of centrifuge model tests to investigate the effect of width of SCP improved ground on the passive earth pressure. They showed that the passive earth pressure of improved ground gradually increased with increasing improvement width.

Mizuno et al. [42] assumed composite ground to be uniform ground having an average undrained shear strength and an internal friction angle, and derived the formula shown in Equations (7.12a) and (7.12b) for predicting the active and passive earth pressures respectively by load equilibrium on an assumed soil block. In this equation, no stress concentration on sand piles was taken into account. They compared the active earth pressure formula with the field measured data acting on the sheet wall during excavation, and showed that the proposed method provided a reasonable estimation of the field data.

$$p_A = \tan^2\left\{\frac{\pi}{4} - \frac{1}{2}\tan^{-1}(as \cdot \tan\phi_s)\right\} \cdot (\gamma \cdot z + q)$$
$$- 2 \cdot (1 - as) \cdot c \cdot \tan\left\{\frac{\pi}{4} - \frac{1}{2}\tan^{-1}(as \cdot \tan\phi_s)\right\}$$

(7.12a)

$$p_P = \tan^2\left\{\frac{\pi}{4} + \frac{1}{2}\tan^{-1}(as \cdot \tan\phi_s)\right\} \cdot \left(\gamma \cdot z + q\right)$$
$$+ 2 \cdot (1 - as) \cdot c \cdot \tan\left\{\frac{\pi}{4} + \frac{1}{2}\tan^{-1}(as \cdot \tan\phi_s)\right\}$$

(7.12b)

where:

- as : replacement area ratio
- c : undrained shear strength of clay (kN/m^2)
- p_A : active earth pressure of improved ground (kN/m^2)
- p_P : passive earth pressure of improved ground (kN/m^2)
- q : overburden pressure (kN/m^2)
- z : depth (m)
- γ : unit weight of improved ground (kN/m^3)
- ϕ_s : internal friction angle of sand pile

7.7   HORIZONTAL RESISTANCE OF IMPROVED GROUND

For this subject, some laboratory, centrifuge and field loading tests were carried out to demonstrate the effectiveness of SCP improvement on the horizontal resistance of a pile or sheet wall. Mikasa et al. [68,73] performed a series of centrifuge model tests on the stability of a sheet wall revetment subjected to back filling.

Higashigawa et al. [92] carried out field loading tests of a steel sheet pile cellular bulkhead on SCP improved ground. Oe et al. [100] conducted field loading tests on the horizontal resistance of SCP improved ground with the replacement area ratio of 0.7 where a single pile and sheet wall were subjected to horizontal loading. They found that the coefficient of subgrade reaction back calculated by the measured load displacement relation decreased with the square root of displacement and that the coefficient had a close relationship with the SPT N-value of sand piles.

Kitazume and Murakami [101] and Kitazume and Miyajima [149] performed a series of centrifuge model tests on sheet pile walls in SCP improved ground subjected to backfilling. They observed the failure mode of sand piles in detail and improvement mechanism of SCP improvement.

The horizontal resistance of SCP improved ground has been evaluated by subgrade reaction theory. The magnitude of coefficient of subgrade reaction of three-dimensional improved ground, composite ground consisting of clay and sand piles, has been studied. Tanigawa et al. [102] and Tanigawa and Sawaguchi [109] performed laboratory model tests on the horizontal resistance of a pile embedded in the improved ground. They proposed that the coefficient of subgrade reaction of improved ground, k_h, was simulated by a parallel connection of the coefficients of subgrade reaction of sand piles and clay as formulated by Equation (7.13).

$$\frac{1}{k_h} = \frac{1 - as}{k_{hc}} + \frac{as}{k_{hs}}$$
(7.13)

where:
    as : replacement area ratio
    k_h : coefficient of subgrade reaction of improved ground (kN/m^3)
    k_{hc} : coefficient of subgrade reaction of clay (kN/m^3)
    k_{hs} : coefficient of subgrade reaction of sand pile (kN/m^3)

Uemura et al. [155] and Ohba [156] introduced the 'improvement index', I_{as}, to evaluate the horizontal resistance of SCP improved ground. They simulated the horizontal resistance by a series connection of springs of sand piles and clay as formulated by Equation (7.14).

$$k_h = (1 - as) \cdot k_{hc} + as \cdot k_{hs}$$
(7.14)

where:
    as : replacement area ratio
    k_h : coefficient of subgrade reaction of improved ground (kN/m^3)
    k_{hc} : coefficient of subgrade reaction of clay (kN/m^3)
    k_{hs} : coefficient of subgrade reaction of sand pile (kN/m^3)

Kitazume et al. [166] performed a series of centrifuge model tests and numerical analyses on the horizontal resistance of a pile in SCP improved ground and found that the coefficient of subgrade reaction of improved ground increased almost linearly with increasing replacement area ratio. They derived the relationship between the coefficient of subgrade reaction and the width of improved ground and showed a design evaluation of SCP improved ground based on the subgrade reaction theory.

Figure 7.9 shows the relationship between the coefficient of subgrade reaction of improved ground and replacement area ratio, as. In the figure, the relationships calculated by Equations (7.13) and (7.14) are plotted as full and broken lines respectively. The figure shows that the parallel connection concept gives a smaller k_h estimation than the series connection concept. In the figure, measured data in laboratory model tests are also plotted together, in which the measured values are normalized with

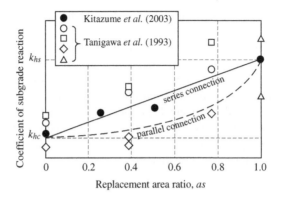

Figure 7.9.   Relationship between coefficient of subgrade reaction of improved ground and replacement area ratio.

respect to the k_{hc} and k_{hs} values for each test [166,102]. As there is much scatter in the measured data probably due to model test conditions, model ground preparation and calculation manner, the difference in the value of k_h estimated by the two concepts is small compared with the scatter of measured data. Further research is necessary on this subject.

7.8   GROUND HEAVING DUE TO INSTALLATION OF SAND PILES

7.8.1   Shape of upheaval ground

Ground heaving during SCP execution has often been encountered. Several researches have been done to investigate the shape and volume of the upheaval portion and the effect of replacement area ratio, type of execution machine, and construction sequence of sand piles.

As the first of these researches, Yabushita et al. [33] conducted field tests to investigate the shape and soil properties of upheaval ground due to sand piles installation. They found that the amount of the upheaval portion was about 80% of the amount of sand installed, and its shape was influenced by the sequence of sand piles installation.

Sogabe [43] accumulated 28 field data on ground upheaval due to sand piles installation, in which the length of SCP piles was less than 20 m. After multiple regression analysis of the field data, he found that the coefficient of upheaval was dominantly influenced by replacement area ratio, length of sand pile and shear strength of the original ground, where the coefficient of upheaval, μ, is defined as the ratio of the volume of the upheaval portion to the volume of sand introduced. According to the analysis, he proposed a method for estimating the coefficient of upheaval, μ, as follows:

$$\mu = 0.316 \cdot as - 0.028 \cdot L + 0.0037 \cdot q_u + 0.700 \tag{7.15}$$

where:

as : replacement area ratio

L : sand pile length (m)

q_u : unconfined compressive strength of original ground at the depth of $L/3$ (ton/m^2)

μ : coefficient of upheaval

In this equation, the effect of SCP pile length is taken into account by a subtraction form, which can make the value of μ negative when the length becomes large.

Shiomi and Kawamoto [70] analyzed another 14 field data together with Sogabe's data to modify the estimation equation. They concluded that the influence of shear strength of the original ground, q_u, on the coefficient of upheaval could be negligible and proposed a new equation to estimate the coefficient of upheaval as follows:

$$\mu = 0.356 \cdot as + 2.803 / L + 0.112 \qquad (7.16)$$

where:

as : replacement area ratio

L : sand pile length (m)

μ : coefficient of upheaval

In this equation, the effect of SCP pile length is taken into account by a reciprocal form instead of a subtraction form to prevent the magnitude of coefficient of upheaval from becoming negative when the sand pile length becomes large. They found the shape of the upheaval portion was also greatly influenced by the sequence of sand piles installation and proposed two methods of estimating the two and three-dimensional shapes of the upheaval portion based on the accumulated field data.

Hirao and Matsuo [64,69] analyzed field data including the field test results at Maizuru Port (see Chapter 5.3), and concluded that Equation (7.16) could be applicable to predict their data. Maeda et al. [81] analyzed field data obtained in the Kansai International Airport construction project and proposed a new method of predicting the shape of the upheaval portion by taking into account the sequence of sand piles installation and the effect of confined condition.

After accumulating field data in the 1980s and 1990s, Hirao [117] found that the field measured data for the coefficient of upheaval in recent field experience were higher than the data estimated by Equation (7.16). He analyzed the field data and found that the sand pile diameter also influenced the coefficient. According to the analysis, he proposed a new equation for estimating the coefficient of the upheaval portion, as follows:

$$\mu = 0.400 \cdot as + 2.477 / L + 0.101 \cdot D + 0.011 \qquad (7.17)$$

where:

D : diameter of sand pile (m)

He assumed that the shape of the upheaval portion was linear and proposed a formula for estimating the shape of upheaval ground according to the accumulated field data.

7.8.2   Shear strength reduction and recovery

Murayama [5] provided a basic concept on the soil disturbance due to sand piles installation and the strength recovery.

Ogawa and Ichimoto [6] investigated the influence of soil disturbance on shear strength.

Enokido *et al.* [32] showed field measurements of shear strength and studied the magnitude and speed of strength recovery. They concluded that the shear strength increase was caused by the increase of effective consolidation pressure due to sand piles installation.

Yabushita *et al.* [33] showed field measurements on shear strength reduction and recovery in marine construction.

Akagi [40] discussed the shear strength reduction and recovery based on laboratory tests and field measurements.

Ichimoto [44] accumulated field data on shear strength reduction and recovery of clay ground due to sand piles installation. He concluded that the shear strength of clay decreased due to sand piles installation but soon recovered within a couple of months.

Sogabe [43] discussed the shear strength at the upheaval portion of clay ground.

Hirao and Matsuo [64] discussed the soil characteristics of clay and found that the unit weight and shear strength of clay increased after the sand piles installation due to consolidation. They also found that the permeability of sand piles was important for accelerating the dissipation of pore water pressure generated during sand piles installation.

Other field measurements were reported which concluded that the shear strength of clay ground decreased considerably after sand piles installation but recovered and increased more than the original strength within a couple of months [74].

Kojima *et al.* [103] measured field data on the shear strength of clay subjected to disturbance and recovery. They found that the shear strength decreased considerably due to sand piles installation, but recovered and increased more than the original strength after a couple of months.

7.9   DENSIFICATION OF SANDY GROUND

7.9.1   Effect of improvement

Regarding the extent of improvement effect of sand piles installation, Nishida and Hoda [3] performed theoretical studies on the expansion mechanism of sandy ground by installation of sand piles and concluded that the influence of sand piles installation was extended to the surroundings up to a diameter of 8 times the sand pile diameter.

Nakayama *et al.* (1973) studied some accumulated SPT *N*-values before and after improvement. They found that the improvement effect was relatively small at a shallow depth of less than 2 m and it was dependent upon the sand/gravel contents of the original ground.

Regarding the effect of fines content, Fudo Construction Co., Ltd. [27] accumulated field SPT N-value data and found that the improvement effect decreased in the case that the fines content of the original ground exceeded about 20%. Similar phenomena were confirmed by Tokimatsu and Yoshimi [174] and Sugiyama et al. [65].

Yamamoto et al. [127] measured field data on the increase in density of sandy ground directly due to sand piles installation, and found that the volumetric strain due to sand piles installation was smaller than the replacement area ratio.

Harada et al. [135] explained that this discrepancy was due to the effect of ground upheaval and horizontal deformation of ground during sand piles installation.

Regarding the effect of horizontal stress increase on liquefaction resistance, Kimura et al. [71,75] measured the horizontal stress by means of borehole horizontal loading tests after sand piles installation. They found that the static earth pressure coefficient, K_0, increased due to sand piles installation and remained high for 21 months after installation. They discussed the possibility of taking into account this effect in the liquefaction design. Similar phenomena were found in laboratory model tests [170,175].

7.9.2   Design procedure

Ogawa and Ishidou [9] proposed a design procedure. In the procedure, the void ratios of original and improved grounds were estimated by the SPT N-value where the effect of uniformity coefficient, U_c, and overburden pressure were taken into account. However, the effects of upheaval and horizontal deformation of ground due to sand piles installation were not taken into account in their procedure.

Fudo Construction Co., Ltd. [27] proposed a design method based on accumulated field data on the relationship between void ratio, relative density and SPT N-value. In the method, the relative density was calculated by the SPT N-value taking into account the maximum and minimum void ratios of the soil, and the effect of overburden pressure was incorporated in the SPT N-value to calculate the void ratio. Due to its simplicity, this design method has frequently been adopted to sandy ground whose fines content is less than 20%.

According to field measurements on the effect of fines content, Mizuno et al. [76] modified Ogawa and Ishidou's procedure [9] to take into account the effect of fines content. In the procedure, the relationship between e_{max} and e_{min} and fines content, presented by Hirama [176] was used to estimate the void ratio of the original ground. The effect of relative density on the SPT N-value was also incorporated by Meyerhof's equation [177]. They introduced a reduction factor to incorporate the effect of fines content on the SPT N-value.

Yamamoto et al. [128,150] proposed a design method in which the effect of ground heave was taken into account. In their proposed method, a new parameter, 'effective compaction factor', Rc, was introduced, which modified the void ratio change by fines content. They proposed the relationship between the relative density and SPT N-value and overburden pressure. This design method was confirmed by field data [158].

Recently, Yamazaki *et al.* [159–161,171] proposed a design method to take into account the effect of fines content on improvement effect.

REFERENCES

1   Murayama, S.: Soft soil improvement by sand compaction pile method (compozer method). Seminar for Osaka Construction Association, pp.1–11, 1957 (in Japanese).

2   Murayama, S.: Soil improvement by sand compaction piles (compozer method). Seminar for Japan Construction Association, pp.1–4, 1958 (in Japanese).

3   Nishida, Y. and Hoda, I.: Fundamental study on compacted sand pile. Journal of Geotechnical Engineering, Japan Society of Civil Engineers, No.69/3, pp.38–44, 1960 (in Japanese).

4   Ogawa, M.: Development of 'compozer method'. Journal of Construction Machinery, No.150, pp.16–19, 1962 (in Japanese).

5   Murayama, S.: Considerations of vibro-compozer method for application to cohesive ground. Journal of Construction Machinery, No.150, pp.10–15, 1962 (in Japanese).

6   Ogawa, M. and Ichimoto, E.: Application of vibro-compozer method to cohesive ground. Journal of the Japanese Society of Soil Mechanics and Foundation Engineering, 'Tsuchi-to-Kiso', Vol.11, No.3, pp.3–9, 1963 (in Japanese).

7   Motouchi, S., Tamaoki, A. and Kozawa, Y.: Bearing capacity of composite ground. Proc. of the 19th Annual Conference of the Japan Society of Civil Engineers, III, pp.64-1-2, 1964 (in Japanese).

8   Ibaraki, T.: Experimental Investigations of direct shear tests of composite soil (1st report). Journal of the Japanese Society of Soil Mechanics and Foundation Engineering, 'Tsuchi-to-Kiso', Vol.13, No.3, pp.19–24, 1965 (in Japanese).

9   Ogawa, M. and Ishidou, M.: Application of 'compozer method' for sandy ground. Journal of the Japanese Society of Soil Mechanics and Foundation Engineering, 'Tsuchi-to-Kiso', Vol.13, No.2, pp.77–81, 1965 (in Japanese).

10   Niimi, Y., Ogawa, H. and Ichimoto, E.: Bearing capacity of sand pile in clay ground. Proc. of the 1st Annual Conference of the Japanese Society of Soil Mechanics and Foundation Engineering, pp.197–202, 1966 (in Japanese).

11   Taguchi, S.: Design chart for slip circle analysis – for Vibro-Compozer method–. Journal of the Japanese Society of Soil Mechanics and Foundation Engineering, 'Tsuchi-to-Kiso', Vol.14, No.5, pp.32–38, 1966 (in Japanese).

12   Ogawa, M. and Koike, H.: Experiment on new replacement method of soft ground. Proc. of the 21st Annual Conference of the Japan Society of Civil Engineers, III, pp.120-1-2, 1966 (in Japanese).

13 Matsuo, M., Kuga, T. and Maekawa, I.: The study on consolidation and shear strength of cohesive soil containing sand pile. Journal of Geotechnical Engineering, Japan Society of Civil Engineers, No.141/3, pp.42–55, 1967 (in Japanese).

14 Matsuo, M.: Stability calculations of clay ground with sand piles. Journal of the Japanese Society of Soil Mechanics and Foundation Engineering, 'Tsuchi-to-Kiso', Vol.15, No.2, pp.27–35, 1967 (in Japanese).

15 Matsuo, M., Inada, N. and Teramura, M.: Studies on bearing capacity of composite ground (1st report). Journal of the Japanese Society of Soil Mechanics and Foundation Engineering, 'Tsuchi-to-Kiso', Vol.16, No.12, pp.21–28, 1968 (in Japanese).

16 Mogami, T., Nakayama, J., Ueda, S., Kuwata, H., Kamata, H. and Taguchi, S.: Laboratory model tests on composite ground (1st report). Journal of the Japanese Society of Soil Mechanics and Foundation Engineering, 'Tsuchi-to-Kiso', Vol.16, No.8, pp.9–17, 1968 (in Japanese).

17 Mogami, T., Nakayama, J., Ueda, S., Kuwata, H., Kamata, H. and Taguchi, S.: Laboratory model tests on composite ground (2nd report). Journal of the Japanese Society of Soil Mechanics and Foundation Engineering, 'Tsuchi-to-Kiso', Vol.16, No.11, pp.5–11, 1968 (in Japanese).

18 Nakayama, J.: Design charts of safety factor against failure of embankment. Journal of the Japanese Society of Soil Mechanics and Foundation Engineering, 'Tsuchi-to-Kiso', Vol.16, No.1, pp.25–31, 1968 (in Japanese).

19 Kuno, G. and Nakayama, J.: The improvement effect of 'compozer method' to stability of embankment. Journal of the Japanese Society of Soil Mechanics and Foundation Engineering, 'Tsuchi-to-Kiso', Vol.16, No.12, pp.11–19, 1968 (in Japanese).

20 Matsuo, M., Teramura, M., Inada, N. and Hirose, T.: Studies on bearing capacity of composite ground (2nd Report). Journal of the Japanese Society of Soil Mechanics and Foundation Engineering, 'Tsuchi-to-Kiso', Vol.17, No.1, pp.3–9, 1969 (in Japanese).

21 Matsuo, M.: Studies on bearing capacity of composite ground (3rd Report). Journal of the Japanese Society of Soil Mechanics and Foundation Engineering, 'Tsuchi-to-Kiso', Vol.17, No.2, pp.5–11, 1969 (in Japanese).

22 Matsuo, M., Nishikawa, M. and Kojima, A.: On the stress distribution in composite ground. Proc. of the Kansai Regional Conference of the Japan Society of Civil Engineering, III, pp.22(1) – (2), 1969 (in Japanese).

23 Aboshi, H., Yoshikuni, H. and Harada, K.: K_0 consolidation of clay ground composed of large diameter sand pile. Proc. of the 5th Annual Conference of the Japanese Society of Soil Mechanics and Foundation Engineering, pp.397–400, 1970 (in Japanese).

24 Aboshi, H., Yoshikuni, H., Ichimoto, E. and Harada, K.: Settlement characteristics of composite ground. Proc. of the 15th Soil Mechanics and Foundation Engineering, pp.73–80, 1970 (in Japanese).

25 Yoshikuni, H. and Ueno, T.: Consolidation behavior vertical drain with stress concentration effect. Proc. of the 6th Annual Conference of the Japanese Society of Soil Mechanics and Foundation Engineering, pp.335–338, 1971 (in Japanese).

26 Matsuo, M., Kuroda, K. and Ibuki, N.: Characteristics of stress distribution in composite ground. Proc. of the Kansai Regional Conference of the Japan Society of Civil Engineering, III, pp.31(1)–(2), 1971 (in Japanese).

27 Fudo Construction Co., Ltd.: Design manual for 'compozer system'. Fudo Construction Co., Ltd., 1971 (in Japanese).

28 Nakayama, J., Ichimoto, E. and Ueda, S.: Improvement effect of slip failure of embankment on improved ground by vibro-compozer method. Proc. of the 15th Soil Mechanics and Foundation Engineering, pp.81–86, 1971 (in Japanese).

29 Murayama, S., Suematsu, N. and Matsuoka, H.: Analyses of composite ground by consideration of stress strain characteristics of sand column. Proc. of the 7th Annual Conference of the Japanese Society of Soil Mechanics and Foundation Engineering, pp.399–402, 1972 (in Japanese).

30 Murayama, S., Matsuoka, H. and Kamo, I.: Analyses of composite ground by consideration of stress strain characteristics of sand column (2nd report). Proc. of the 8th Annual Conference of the Japanese Society of Soil Mechanics and Foundation Engineering, pp.407–410, 1973 (in Japanese).

31 Matsuoka, T., Tsukai, N. and Harada, K.: Theory and actual behavior of soft soil improvement (compozer method). Journal of the Japanese Society of Soil Mechanics and Foundation Engineering, 'Tsuchi-to-Kiso', Vol.21, No.11, pp.41–46, 1973 (in Japanese).

32 Enokido, M., Takahashi, Y., Gotoh, S. and Maeda, K.: Strength recovery of cohesive ground due to sand piles installation. Journal of the Japanese Society of Soil Mechanics and Foundation Engineering, 'Tsuchi-to-Kiso', Vol.21, No.6, pp.87–92, 1973 (in Japanese).

33 Yabushita, H., Terada, M. and Suematsu, N.: Compulsory replacement method of soft clay ground. Journal of the Japanese Society of Soil Mechanics and Foundation Engineering, 'Tsuchi-to-Kiso', Vol.21, No.9, pp.33–40, 1973 (in Japanese).

34 Nakayama, J., Uchimoto, E., Kamada, H. and Taguchi, S.: On stabilization characteristics of sand compaction piles. SOILS AND FOUNDATIONS, Vol.13, No.3, pp.61–68, 1973.

35 Tominaga, M., Hashimoto, M. and Mizuno, Y.: Slag compaction pile method. Proc. of the 9th Annual Conference of the Japanese Society of Soil Mechanics and Foundation Engineering, pp.901–904, 1974 (in Japanese).

36 Murayama, S., Suematsu, N. and Iwasaki, M.: Analyses of composite ground by consideration of stress strain characteristics of sand column (3rd report). Proc. of the 9th Annual Conference of the Japanese Society of Soil Mechanics and Foundation Engineering, pp.449–452, 1974 (in Japanese).

37 Yabushita, H., Suematsu, N., Tsuboi, H. and Kanda, Y.: Influence of design parameters on safety factor of slip circle analysis based on practices. Proc. of the

10th Annual Conference of the Japanese Society of Soil Mechanics and Foundation Engineering, pp.823–826, 1975 (in Japanese).

38 Yabushita, H., Mizuno, Y., Iwasaki, M. and Sawai, M.: Consideration on earth pressure of composite ground. Proc. of the 11th Annual Conference of the Japanese Society of Soil Mechanics and Foundation Engineering, pp.717–720, 1976 (in Japanese).

39 Yamaguchi, H. and Murakami, Y.: Stress concentration ratio of composite ground. Proc. of the 12th Annual Conference of the Japanese Society of Soil Mechanics and Foundation Engineering, pp.543–546, 1977 (in Japanese).

40 Akagi, T.: Strength increase of soft clay due to sand piles installation. Proc. of the 12th Annual Conference of the Japanese Society of Soil Mechanics and Foundation Engineering, pp.1241–1244, 1977 (in Japanese).

41 Yoshikuni, H.: Design and quality control of vertical drain method. Gihodo Shuppan Co., Ltd., pp.208, 1979 (in Japanese).

42 Mizuno, Y., Yoshida, M. and Tsuboi, H.: Estimation of earth pressure of composite ground. Proc. of the 34th Annual Conference of the Japan Society of Civil Engineers, III, pp.395–396, 1979 (in Japanese).

43 Sogabe, T.: Technical subjects on design and execution of sand compaction pile method. Proc. of the 36th Annual Conference of the Japan Society of Civil Engineers, III, pp.39–50, 1981 (in Japanese).

44 Ichimoto, E.: Practical design method and calculation of sand compaction pile method. Proc. of the 36th Annual Conference of the Japan Society of Civil Engineers, III, pp.51–55, 1981 (in Japanese).

45 Takada, N. and Nogawa, H.: Compressibility of clay containing sand-compaction piles. Journal of the Japanese Society of Soil Mechanics and Foundation Engineering, 'Tsuchi-to-Kiso', Vol.30, No.2, pp.47–54, 1982 (in Japanese).

46 Matsuo, M. and Suzuki, H.: Study on reliability-based design of improvement of clay layer by sand compaction piles. SOILS AND FOUNDATIONS, Vol.23, No.3, pp.112–122, 1983.

47 Kusano, I.: Actual practices and subjects of sand compaction pile method (3) – improvement effect of river dike and dynamic response –. Journal of the Japanese Society of Soil Mechanics and Foundation Engineering, 'Tsuchi-to-Kiso', Vol.31, No.4, pp.79–85, 1983 (in Japanese).

48 Ichimoto, E. and Suematsu, N.: Actual practices and subjects of sand compaction pile method (3) – summary –. Journal of the Japanese Society of Soil Mechanics and Foundation Engineering, 'Tsuchi-to-Kiso', Vol.31, No.5, pp.83–90, 1983 (in Japanese).

49 Takechi, O., Yagi, N. and Yatabe, R.: On the stress distribution in composite ground during consolidation. Proc. of the 50th Annual Conference of the Japan Society of Civil Engineers, III, pp.449–450, 1983 (in Japanese).

50 Yoshikuni, H., Mae, K. and Matsukata, K.: Undrained triaxial compression tests of clay with a sand pile. Proc. of the Symposium on Strength and Deformation of Composite Ground, pp.119–124, 1984 (in Japanese).

51   Yagi, N., Enoki, M. and Yatabe, R.: Mechanical properties of composite ground on model tests. Proc. of the Symposium on Strength and Deformation of Composite Ground, pp.147–152, 1984 (in Japanese).

52   Saitou, K., Kusakabe, O., Kimura, T. and Nakase, A.: On the mechanical behaviour of soft clay improved by sand compaction piles. Proc. of the Symposium on Strength and Deformation of Composite Ground, pp.107–112, 1984 (in Japanese).

53   Mikasa, M., Takada, N., Oshima, A., Kawamoto, K. and Higashi, S.: An experimental study on the stability of clay ground improved by sand compaction piles. Proc. of the Symposium on Strength and Deformation of Composite Ground, pp.113–118, 1984 (in Japanese).

54   Inoue, T., Mukai, M. and Koba, Z.: Deformation characteristics of ground improved by SCP with low sand-replacement ratio. Proc. of the Symposium on Strength and Deformation of Composite Ground, pp.129–134, 1984 (in Japanese).

55   Kurihara, N. and Tochigi, H.: Consideration on the deformation characteristics of sand compaction pile method with field measurement. Proc. of the Symposium on Strength and Deformation of Composite Ground, pp.135–138, 1984 (in Japanese).

56   Fukuoka, M., Sago, J., Taya, M., Higuchi, K., Kusakabe, F. and Kawanabe, S.: The effect of the soil improvement by sand compaction pile method. Proc. of the Symposium on Strength and Deformation of Composite Ground, pp.153–158, 1984 (in Japanese).

57   Nakanodou, H., Moriwaki, T. and Yamamoto, M.: Effect of improvement area on bearing capacity of SCP improved ground. Proc. of the Symposium on Strength and Deformation of Composite Ground, pp.159–164, 1984 (in Japanese).

58   Yagi, N., Enoki, M., Yatabe, R., Takechi, O. and Kumamoto, N.: Characteristics of composite ground during consolidation. Proc. of the 19th Annual Conference of the Japanese Society of Soil Mechanics and Foundation Engineering, pp.1555–1556, 1984 (in Japanese).

59   Suematsu, N.: Sand compaction pile method – state of art report–. Proc. of the Symposium on Strength and Deformation of Composite Ground, pp.13–26, 1984 (in Japanese).

60   Hirao, H.: Experimental study on stress ratio of compacted sand pile. Proc. of the Symposium on Strength and Deformation of Composite Ground, pp.125–128, 1984 (in Japanese).

61   Fukumoto, K. and Yamagata, K.: On the stresses supported by a sand pile and clay in the composite ground. Proc. of the Symposium on Strength and Deformation of Composite Ground, pp.139–146, 1984 (in Japanese).

62   Suematsu, N. and Tsuboi, H.: Consideration of stress concentration ratio on a composite ground. Proc. of the Symposium on Strength and Deformation of Composite Ground, pp.165–170, 1984 (in Japanese).

63   Kimura, T., Nakase, A., Kusakabe, O. and Saitoh, K.: Behaviour of soil improved by sand compaction piles. Proc. of the 11th International Conference on Soil Mechanics and Foundation Engineering, pp.1109–1112, 1985.

64 Hirao, H. and Matsuo, M.: Study on upheaval ground generated by sand com-
    paction piles. Journal of Geotechnical Engineering, Japan Society of Civil
    Engineers, No.364/3-4, pp.169–178, 1985 (in Japanese).
65 Sugiyama, S., Fujisaku, T., Tokitoh, M. and Taniguchi, D.: A case study of vibro-
    compaction on sandy ground containing fines. Journal of the Japanese Society of
    Soil Mechanics and Foundation Engineering, 'Tsuchi-to-Kiso', Vol.33, No.4,
    pp.43–47, 1985 (in Japanese).
66 Enoki, M., Yagi, N. and Yatabe, R.: Stress concentration on sand column during
    one-dimensional consolidation of composite ground. Journal of Geotechnical
    Engineering, Japan Society of Civil Engineers, No.376/3-6, pp.201–209, 1986 (in
    Japanese).
67 Fujii, S. and Kobayashi, N.: An example of measurement on soft ground. Proc. of
    the 21st Annual Conference of the Japanese Society of Soil Mechanics and
    Foundation Engineering, pp.1859–1860, 1986 (in Japanese).
68 Mikasa, M., Takada, N., Ohshima, A., Kawamoto, K., Higashi, S., and Ohbayashi,
    J.: Centrifuged model tests on sheet pile quay wall on clay ground improved by sand
    compaction piles. Proc. of the 21st Annual Conference of the Japanese Society of
    Soil Mechanics and Foundation Engineering, pp.1857–1858, 1986 (in Japanese).
69 Hirao, H. and Matsuo, M.: Study on characteristics of upheaval part of cohesive
    ground caused by soil improvement. Journal of Geotechnical Engineering, Japan
    Society of Civil Engineers, No.376/3-6, pp.277–285, 1986 (in Japanese).
70 Shiomi, M. and Kawamoto, K.: Prediction of ground heave associated with the
    installation of sand compaction piles. Proc. of the 21st Annual Conference of the
    Japanese Society of Soil Mechanics and Foundation Engineering, pp.1861–1862,
    1986 (in Japanese).
71 Kimura, T., Okumura, I., Misawa, H. and Kawanabe, O.: Variation lateral stress
    ratio for one-dimensional strain. Proc. of the 21st Annual Conference of the
    Japanese Society of Soil Mechanics and Foundation Engineering, pp.1863–1866,
    1986 (in Japanese).
72 Okamoto, T., Ito, K., Nakajima, Y. and Okita, Y.: An executive example of gravel
    drain system in urban area. Proc. of the 21st Annual Conference of the Japanese
    Society of Soil Mechanics and Foundation Engineering, pp.1867–1868, 1986 (in
    Japanese).
73 Mikasa, M., Takada, N., Kurisu, K. and Oobayashi, J.: Centrifuge model tests on
    sheet pile quay wall on clay ground improved by sand compaction piles (2nd
    report). Proc. of the 22nd Annual Conference of the Japanese Society of Soil
    Mechanics and Foundation Engineering, pp.1841–1842, 1987 (in Japanese).
74 Enoki, M., Yagi, N., Yatabe, R., Tasaka, Y. and Takada, T.: Consideration in initial
    stress state of composite ground. Proc. of the 22nd Annual Conference of the
    Japanese Society of Soil Mechanics and Foundation Engineering, pp.1801–1802,
    1987 (in Japanese).
75 Kimura, T., Shiota, K., Misawa, H. and Kawanabe, O.: Variation lateral stress
    ratio for one-dimensional strain (2nd report). Proc. of the 22nd Annual

Conference of the Japanese Society of Soil Mechanics and Foundation Engineering, pp.1795–1796, 1987 (in Japanese).

76 Mizuno, Y., Suematsu, N. and Okuyama, K.: Design method for sand compaction pile for sandy soils containing fines. Journal of the Japanese Society of Soil Mechanics and Foundation Engineering, 'Tsuchi-to-Kiso', Vol.35, No.5, pp.21–26, 1987 (in Japanese).

77 Shoji, A., Morisaki, K., Watanabe, K. and Goami, Y.: Automatic control system of SCP bessel. Journal of the Foundation Engineering & Equipment, 'Kisoko', Sogodoboku, No.4, pp.88–92, 1988 (in Japanese).

78 Kitazume, M., Minami, K., Matsui, H. and Naruse, E.: Field test on applicability of copper slag sand to sand compaction pile method. Proc. of the 3rd International Congress on Environmental, Vol.2, pp.643–648, 1988.

79 Fukute, T., Higuchi, Y., Furuichi, M. and Tsuboi, H.: Prediction on the shape of upheaved ground caused by sand compaction piles. Proc. of the 33rd Symposium of the Japanese Society of Soil Mechanics and Foundation Engineering, pp.23–28, 1988 (in Japanese).

80 Okada, Y., Yagyuu, T. and Sawada, Y.: Field Loading test of SCP improved ground with low replacement area ratio. Journal of the Japanese Society of Soil Mechanics and Foundation Engineering, 'Tsuchi-to-Kiso', Vol.37, No.8, pp.57–62, 1989 (in Japanese).

81 Maeda, S., Takai, T. and Fukute, T.: Shape and properties of the upheaval of cohesive soil improved by compacted sand piling method. Journal of Construction Management and Engineering, Japan Society of Civil Engineers, No.403/6–10, pp.55–63, 1989 (in Japanese).

82 Enoki, M., Yagi, N., Yatabe, R. and Ichimoto, E.: Shearing Characteristic of composite ground and its application to stability analysis. Deep Foundation Improvements: Design, Construction, and Testing, ASTM STP 1089, Robert C. Bachus, Ed., American Society for Testing and Materials, pp.19–31, 1990.

83 Aboshi, H., Mizuno, Y. and Kuwabara, M.: Present state of sand compaction pile in Japan. Deep Foundation Improvements: Design, Construction, and Testing, ASTM STP 1089, Robert C. Bachus, Ed., American Society for Testing and Materials, pp.32–46, 1990.

84 Terashi, M., Kitazume, M. and Minagawa, S.: Bearing capacity of improved ground by sand compaction piles. Deep Foundation Improvements: Design, Construction, and Testing, ASTM STP 1089, Robert C. Bachus, Ed., American Society for Testing and Materials, pp.47–61, 1990.

85 Terashi, M. and Kitazume, M.: Bearing capacity of clay ground improved by sand compaction piles of low replacement area ratio. Report of the Port and Harbour Research Institute, Vol.29, No.2, pp.119–148, 1990 (in Japanese).

86 Takai, T., Imano, K., Ogino, H. and Nakamura, M.: A study of cellular shell driving into ground using vibration hammers. Journal of Construction Management and Engineering, Japan Society of Civil Engineers, No.415/6-12, pp.53–62, 1990 (in Japanese).

87 Kanda, K. and Terashi, M.: Practical formula for the composite ground improved by sand compaction pile method. Technical Note of the Port and Harbour Research Institute, No.669, pp.52, 1990 (in Japanese).

88 Asaoka, A., Matsuo, M. and Kodaka, T.: Undrained bearing capacity of clay with sand piles. Proc. of the 9th Asian Regional Conference on Soil Mechanics and Foundation Engineering, pp.467–470, 1991.

89 Ichimoto, E.: Study on stability analysis of composite ground. Ph.D. Theses, 1991 (in Japanese).

90 Shinsha, H., Takata, K., Kurumada, Y. and Fujii, N.: Centrifuge model tests on clay ground partly improved by sand compaction piles. Proc. of the International Conference on Centrifuge Modeling, CENTRIFUGE 91, pp.311–318, 1991.

91 Takemura, J., Watabe, Y., Suemasa, N., Hirooka, A. and Kimura, T.: Stability of soft clay improved with sand compaction piles. Proc. of the 9th Asian Regional Conference on Soil Mechanics and Foundation Engineering, pp.543–546, 1991.

92 Higashigawa, T., Achiwa, F., Sunami, S. and Matsuo, M.: Study on mechanical behavior of the deep embedded steel sheet pile cellular bulkhead. Journal of Construction Management and Engineering, Japan Society of Civil Engineers, No.435/6-15, pp.95–102, 1991 (in Japanese).

93 Terashi, M., Kitazume, M. and Okada, H.: Applicability of the practical formula for bearing capacity of clay improved by SCP. Proc. of the International Conference on Geotechnical Engineering for Coastal Development, Geo-Coast '91, Vol.1, pp.405–410, 1991.

94 Asaoka, A., Kodaka, T. and Matsuo, M.: The study on undrained bearing capacity of composite ground. Journal of Geotechnical Engineering, Japan Society of Civil Engineers, No.448/3-19, pp.63–71, 1992 (in Japanese).

95 Ishizaki, H.: Deformation analysis of composite ground under undrained condition. Journal of Geotechnical Engineering, Japan Society of Civil Engineers, No.448/3-19, pp.53–62, 1992 (in Japanese).

96 Terashi, M., Kitazume, M. and Kubo, S.: Centrifuge modeling on passive earth pressure of improved ground by sand compaction pile method. Proc. of the 27th Annual Conference of the Japanese Society of Soil Mechanics and Foundation Engineering, pp, 2159–2162, 1992 (in Japanese).

97 Hashidate,Y., Fukuda, T., Okumura, T. and Kobayashi, M.: Engineering properties of oyster shell-sand mixtures. Proc. of the 28th Annual Conference of the Japanese Society of Soil Mechanics and Foundation Engineering, pp.869–872, 1993 (in Japanese).

98 Terashi, M., Kitazume, M., Owaki, T. and Okada, H.: Stability of clay improved by sand compaction pile method. Soil reinforcement: full scale experiments of the 80's, pp.607–634, 1993.

99 Okada, Y.: Ground deformation and strength of improved ground by SCP method with low replacement area ratio. Ph.D. Theses, 1993 (in Japanese).

100 Oe, M., Owaki, T. and Ojima, R.: Design and execution of double sheet walls in Amagasaki Port. Journal of the Foundation Engineering & Equipment, 'Kisoko', Sogodoboku, No.11, pp.89–95, 1993 (in Japanese).

101 Kitazume, M. and Murakami, K.: Behavior of sheet pile walls in the improved ground by sand compaction piles of low replacement area ratio. Report of the Port and Harbour Research Institute, Vol.32, No.2, pp.183–211, 1993 (in Japanese).

102 Tanigawa, M., Sawaguchi, M. and Tanaka, M.: Horizontal resistance of pile in composite ground – effect of improvement area ratio on horizontal coefficient of subgrade reaction –. Proc. of the 28th Annual Conference of the Japanese Society of Soil Mechanics and Foundation Engineering, pp.1815–1816, 1993 (in Japanese).

103 Ojima, R., Kawai, N., Oe, M., Sakai, A. and Mino, S.: Utilization of upheaved seabed ground generated by sand compaction piles. Proc. of the 28th Annual Conference of the Japanese Society of Soil Mechanics and Foundation Engineering, pp.2475–2176, 1993 (in Japanese).

104 Fudo Construction Co., Ltd.: Report of disasters by Kushiro-oki earthquake. Technical Report of Fudo Construction Co., Ltd., 1993 (in Japanese).

105 Takahashi, Y., Koyamachi, N., Nitao, H. and Shigeno, M.: Application of sand compaction pile method with high frequency and small diameter. Journal of the Japanese Society of Soil Mechanics and Foundation Engineering, 'Tsuchi-to-Kiso', Vol.42, No.4, pp.45–48, 1994 (in Japanese).

106 Hashidate, Y., Fukuda, T., Okumura, T. and Kobayashi, M.: Engineering properties of oyster shell-sand mixtures and their application to sand compaction piles. Proc. of the 29th Annual Conference of the Japanese Society of Soil Mechanics and Foundation Engineering, pp.717–720, 1994 (in Japanese).

107 Asaoka, A., Kodaka, T. and Nozu, M.: Undrained shear strength of clay improved with sand compaction piles. SOILS AND FOUNDATIONS, Vol.34, No.4, pp.23–32, 1994.

108 Ishizaki, H.: The effect of improvement condition on deformation behavior of composite ground. Journal of Geotechnical Engineering, Japan Society of Civil Engineers, No.487/3–26, pp.89-98, 1994 (in Japanese).

109 Tanigawa, M. and Sawaguchi, M.: Horizontal resistance of pile in composite ground (2) – effect of improvement area ratio on horizontal coefficient of subgrade reaction –. Proc. of the 29th Annual Conference of the Japanese Society of Soil Mechanics and Foundation Engineering, pp.1599–1600, 1994 (in Japanese).

110 Ando, Y., Tsuboi, H., Yamamoto, M., Harada, K. and Nozu, M.: Recent soil improvement methods for preventing liquefaction. Proc. of the 1st International Conference on Earthquake Geotechnical Engineering, pp.1–6, 1995.

111 Sakai, S., Nozu, M. and Shinkawa, N.: Effect of improvement of mini gravel compaction pile method. Proc. of the 30th Annual Conference of the Japanese Society of Soil Mechanics and Foundation Engineering, pp.2163–2164, 1995 (in Japanese).

112 Sakai, S., Ohbayashi, J. and Takano, S.: Effect of the execution of mini gravel compaction pile method on surround ground. Proc. of the 30th Annual Conference of the Japanese Society of Soil Mechanics and Foundation Engineering, pp.2165–2166, 1995 (in Japanese).

113 Nozu, M.: Study on soil improvement mechanism due to sand pile. Ph.D. Theses, 1995 (in Japanese).

114 Matsuo, M., Kimura, M., Nishio, Y. and Ando, H.: Study on development of soil improvement method using construction waste soil. Journal of Geotechnical Engineering, Japan Society of Civil Engineers, No.547/3-36, pp.199–210, 1996 (in Japanese).

115 Kitazume, M., Miyajima, M. and Nishida, Y.: Stability of revetment on soft clay improved by SCP. Proc. of the 2nd International Conference on Soft Soil Engineering, pp.455–460, 1996.

116 Kobayashi, H., Kogo, M., Suzuki, K. and Sakai, S.: Estimation of the clay ground improved by sand compaction piles at Kawasaki man-made island. Journal of Construction Management and Engineering, Japan Society of Civil Engineers, No.553/6-33, pp.41–48, 1996 (in Japanese).

117 Hirao, H.: Characteristics and beneficial usage of upheaved ground due to sand piles installation. Ph.D. Theses, pp.150, 1996 (in Japanese).

118 Ando, Y., Yamamoto, M., Harada, K. and Nozu, M.: Experimental study on densification of loose sandy soils by penetrating sand piles. Proc. of the 31st Annual Conference of the Japanese Society of Soil Mechanics and Foundation Engineering, pp.73–74, 1996 (in Japanese).

119 Suganuma, M., Fukada, H. and Nakai, N.: Case history of non-vibratory sand compaction method. Proc. of the 52nd Annual Conference of the Japan Society of Civil Engineers, III, pp.412–413, 1997 (in Japanese).

120 Itou, R. and Kamei, T.: Alicability of steel slag to soil improvement. Journal of the Japanese Society of Soil Mechanics and Foundation Engineering, 'Tsuchi-to-Kiso', Vol.21, No.12, pp.37–43, 1997 (in Japanese).

121 Oikawa, K., Matsunaga, Y., Takahashi, K. and Hashimoto, T.: Applicability of steelmaking slag and crushed concrete to SCP material. Proc. of the 32nd Annual Conference of the Japanese Society of Soil Mechanics and Foundation Engineering, pp.2331–2332, 1997 (in Japanese).

122 Matsuo, M., Kimura, M., Nishio, Y. and Ando, H.: Development of soil improvement method using construction waste soil. Journal of Construction Management and Engineering, Japan Society of Civil Engineers, No.567/3-35, pp.237–248, 1997 (in Japanese).

123 Minami, K., Matsui, H., Naruse, E. and Kitazume, M.: Field test on sand compaction pile method with copper slag sand. Journal of Construction Management and Engineering, Japan Society of Civil Engineers, No.574/6-36, pp.49–55, 1997 (in Japanese).

124 Kitazume, M., Miyajima, M. and Nishida, Y.: Bearing capacity of SCP improved clay ground under a revetment. Proc. of the 3rd International Conference on Ground Improvement Geosystems, 1997.

125   Kitazume, M., Miyajima, S. and Sugiyama, T.: Passive earth pressure of SCP improved ground. Proc. of the 32nd Annual Conference of the Japanese Geotechnical Society, pp.2325–2326, 1997 (in Japanese).

126   Tsuboi, H.: Study on vibration mechanism of penetrated column and its application. Ph.D. Theses, 1997 (in Japanese).

127   Yamamoto, M., Harada, K., Nozu, M. and Ohbayashi, J.: Study on evaluation of increasing density due to penetrating sand piles. Proc. of the 32nd Annual Conference of the Japanese Society of Soil Mechanics and Foundation Engineering, pp.2631–2632, 1997 (in Japanese).

128   Yamamoto, M., Sakai, S., Nakasumi, I., Higashi, S., Nozu, M. and Suzuki, A.: Estimation of improvement effects on sandy ground by sand compaction pile. Proc. of the 32nd Annual Conference of the Japanese Society of Soil Mechanics and Foundation Engineering, pp.2315–2316, 1997 (in Japanese).

129   Tsuboi, H., Ando, Y., Harada, K., Ohbayashi, J. and Matsui, T.: Development and application of non-vibratory sand compaction pile method. Proc. of the 8th International Offshore and Polar Engineering Conference, pp.615–620, 1998.

130   Nozu, M., Ohbayashi, J. and Matsunaga, Y.: Application of the static sand compaction pile method to loose sandy soil. Proc. of the International Conference on Problematic Soils, pp.751–755, 1998.

131   Waga, A., Asami, Y., Fukada, H., Nakai, N., Harada, K. and Nozu, M.: A consideration about improvement area of non-vibratory sand compaction method (SAVE compozer). Proc. of the 33rd Annual Conference of the Japanese Society of Soil Mechanics and Foundation Engineering, pp.2159–2160, 1998 (in Japanese).

132   Kitazume, M., Shimoda, Y. and Miyajima, S.: Behavior of sand compaction piles constructed from copper slag sand. Proc. of the International Conference on Centrifuge Modeling, CENTRIFUGE 98, 1998.

133   Jung, J.B., Moriwaki, T., Sumioka, N. and Kusakabe, O.: Consolidation behavior of composite ground improved by sand compaction piles. Proc. of the International Conference on Centrifuge Modeling, CENTRIFUGE 98, pp.825–830, 1998.

134   Ishihara, K., Tsukamoto, Y., Satou, M., Harada, K., Yabe, H. and Amamiya, M.: Soil densification due to static sand pile penetration by hollow cylindrical torsional shear tests. Proc. of the 33rd Annual Conference of the Japanese Society of Soil Mechanics and Foundation Engineering, pp.2157–2158, 1998 (in Japanese).

135   Harada, K., Yamamoto, M. and Ohbayashi, J.: On K_0 value of improved ground by static compaction technique. Proc. of the 53rd Annual Conference of the Japan Society of Civil Engineers, III, pp.544–545, 1998 (in Japanese).

136   Ohbayashi, J., Harada, K., Yamamoto, M. and Sasaki, Y.: Evaluation of liquefaction resistance of compacted ground. Proc. of the Symposium on Earthquake Engineering, pp.1411–141416, 1998 (in Japanese).

137   Nakano, K., Taniguchi, B. and Nakai, N.: The application of materials for non-vibratory sand compaction method (SAVE compozer). Proc. of the 34th Annual

Conference of the Japanese Society of Soil Mechanics and Foundation Engineering, pp.1121–1122, 1999 (in Japanese).

138 Jung, J.B., Moriwaki, T., Sumioka, N. and Kusakabe, O.: The consolidation behavior of clay ground improved by partly penetrated SCP. Journal of Geotechnical Engineering, Japan Society of Civil Engineers, No.617/3-46, pp.101–113, 1999 (in Japanese).

139 Oda, K. and Matsui, T.: Stress sharing mechanism of soft clay ground improved by sand compaction piles with low replacement area ratio. Journal of Geotechnical Engineering, Japan Society of Civil Engineers, No.631/3-48, pp.339–353, 1999 (in Japanese).

140 Ministry of Transport: Technical standards and commentaries for port and harbour facilities in Japan. Ministry of Transport, Japan, 1999 (in Japanese).

141 Coastal Development Institute of Technology: Chapter 12, Design of soil improvement. Design case histories of port and harbor facilities, pp.12-1–12-18, 1999 (in Japanese).

142 Yamamoto, M. and Nozu, M.: Effects on environmental aspect of new sand compaction pile method for soft soil. Proc. of Coastal Geotechnical Engineering in Practice, pp.563–568, 2000.

143 Kitazume, M. and Yasuda, T.: Centrifuge model tests on horizontal resistance of pile in SCP improved ground with Ferro-Nickel Slag. Proc. of the 35th Annual Conference of the Japanese Geotechnical Society, pp.1441–1442, 2000 (in Japanese).

144 Fujiwara, T., Tanaka, H., Kadota, M., Ohiso, Y. and Kurata, T.: Field test of static compaction improved method by granulated excavated-soil. Proc. of the 35th Annual Conference of the Japanese Society of Soil Mechanics and Foundation Engineering, pp.1425–1426, 2000 (in Japanese).

145 Ishii, H., Horikoshi, K., Yamaguchi, J., Satoh, S. and Yamamoto, M.: Model tests for ground compaction behavior of new material for static compaction method. Proc. of the 35th Annual Conference of the Japanese Society of Soil Mechanics and Foundation Engineering, pp.1427–1428, 2000 (in Japanese).

146 Okado, M., Ogura, R., Wada, K. and Kobayashi, Y.: The application of recycled materials for static sand compaction pile method. Proc. of the 35th Annual Conference of the Japanese Society of Soil Mechanics and Foundation Engineering, pp.1431–1432, 2000 (in Japanese).

147 Rahman, Z., Takemura, J., Kouda, M. and Yasumoto, K.: Experimental study on deformation of soft clay improved by low replacement ratio SCP under backfilled caisson loading. SOILS AND FOUNDATIONS, Vol.40, No.5, pp.19–35, 2000.

148 Kitazume, M., Miyajima, S. and Yasuda T.: Improvement effect of SCP improvement on horizontal resistance of a pile. Proc. of the 4th Symposium on Soil Improvement, The Society of Materials Science, Japan, pp.63–68, 2000 (in Japanese).

149 Kitazume, M. and Miyajima, S.: Effect of SCP improvement on stability of sheet pile wall. Proc. of the International Conference on Geotechnical & Geological Engineering, 2000.

150 Yamamoto, M., Harada, K. and Nozu, M.: New design of sand compaction pile for preventing liquefaction in loose sandy ground. Journal of the Japanese Society of Soil Mechanics and Foundation Engineering, 'Tsuchi-to-Kiso', Vol.48, No.11, pp.17–20, 2000 (in Japanese).

151 Kitazume, M., Miyajima, S. and Yasuda T.: Applicability of ferro-nickel slag as a sand compaction pile method material. Proc. of the 3rd BGA International Geoenvironmental Engineering Conference, pp.27–32, 2001.

152 Ohbayashi, J.: Study on damage mechanism and countermeasures of embankment due to liquefaction. Ph.D. Theses, 2001 (in Japanese).

153 International Navigation Association: Seismic design guidelines for port structures., 2001.

154 Kakehashi, T., Umeki, Y., Ookori, K. Soaji, Y., Makibe, Y. and Satou, M.: Technological introduction of low vibration and low noise type ground improvement construction method – Ecological Gentle Geoimprovement –. Proc. of the 57th Annual Conference of the Japan Society of Civil Engineers, III, pp.151–152, 2002 (in Japanese).

155 Uemura, R., Ohba, H., Kusakabe, O. and Takemura, J.: Behavior of a pile installed in SCP ground. Proc. of the 37th Annual Conference of the Japanese Society of Soil Mechanics and Foundation Engineering, pp.1085–1086, 2002 (in Japanese).

156 Ohba, H., Uemura, R. and Kusakabe, O.: Quantitative evaluation of soil improvement by SCP. Proc. of the 37th Annual Conference of the Japanese Society of Soil Mechanics and Foundation Engineering, pp.1087–1088, 2002 (in Japanese).

157 Kitazume, M., Sugano, T., Ohbayashi, J., Nishida, N., Ishimaru, I. and Nakayama, Y.: Centrifuge model tests on dynamic properties of sand compaction pile improved ground. Technical Note of the Port and Airport Research Institute, 2002 (in Japanese).

158 Sugaya, Y., Maekawa, Y., Matsumura, E. and Uno, M.: Improvement effect of SAVE compozer, and consideration of the new design to the sandy ground of sand compaction pile. Proc. of the 37th Annual Conference of the Japanese Society of Soil Mechanics and Foundation Engineering, pp.1095–1096, 2002 (in Japanese).

159 Yamazaki, H., Morikawa, Y. and Koike, F.: Study on design method for densification of sandy deposits by sand compaction pile method. Report of the Port and Airport Research Institute, Vol.41, No.2, pp.93–118, 2002 (in Japanese).

160 Yamazaki, H., Morikawa, Y. and Koike, F.: Study on prediction of SPT N-value of sandy deposits improved by sand compaction pile method. Journal of Geotechnical Engineering, Japan Society of Civil Engineers, No.708/3-59, pp.199–210, 2002 (in Japanese).

161 Yamazaki, H., Morikawa, Y. and Koike, F.: Study on effect of fines content and drainage characteristics of sandy deposits on sand compaction pile method. Journal of Geotechnical Engineering, Japan Society of Civil Engineers, No.722/3-61, pp.303–314, 2002 (in Japanese).

162 Kato, M., Tanaka, Y., Ichikawa, H. and Mishiro, N.: Geo-KONG method. Journal of the Foundation Engineering & Equipment, 'Kisoko', Sogodoboku, No.12, pp.38–41, 2003 (in Japanese).

163 Tsuboi, H., Harada, K., Tanaka, Y. and Matsui, T.: Considerations for judging the suitability of filling materials for an SCP. International Journal of Offshore and Polar Engineering, Vol.13, No.1, 2003.

164 Nishida, N., Watanabe, N., Hattori, M. and Shinoi, T.: Development of sand compaction pile method using inner-screw and high pressure intermittent air. (1) –abstract. Proc. of the 38th Annual Conference of the Japanese Society of Soil Mechanics and Foundation Engineering, pp.1057–1058, 2003 (in Japanese).

165 Otsuka, M. and Isoya, S.: SAVE marine method. Journal of Construction Machine, No.8, pp.58–61, 2003 (in Japanese).

166 Kitazume, M., Takahashi, H. and Takemura, S.: Characteristics of horizontal resistance by sand compaction pile improved ground. Report of the Port and Airport Research Institute, 2003 (in Japanese).

167 Kikuchi, Y.: Application of SPT N-value in Technical Standards for Port and Harbour Facilities. Journal of the Foundation Engineering & Equipment, 'Kisoko', Sogodoboku, No.2, pp.31–35, 2003 (in Japanese).

168 Sugano, T., Kitazume, M., Nakayama, Y., Kawamata, Y., Oobayashi, J., Nishida, N. and Ishimaru, I.: A study on dynamic properties of sand compaction pile improved ground. Technical Note of the Port and Airport Research Institute, No.1047, 32p., 2003 (in Japanese).

169 Okamura, M., Ishihara, M. and Oshita, T.: Liquefaction resistance of sand deposit improved with sand compaction piles. SOILS AND FOUNDATIONS, Vol.43, No.5, pp.175–187, 2003.

170 Satoh, T.: The study of the strength behavior in the ground during rod penetrate and sand pile expand for compaction. Proc. of the 38th Annual Conference of the Japanese Society of Soil Mechanics and Foundation Engineering, pp.1071–1072, 2003 (in Japanese).

171 Yamazaki, H., Morikawa, Y. and Koike, F.: Study on effect of K_0-value on SPT N-value prediction after densification by sand compaction pile method. Journal of Geotechnical Engineering, Japan Society of Civil Engineers, No.750/3-65, pp.231–236, 2003 (in Japanese).

172 Hughes, J.M.O. and Withers, N.J.: Reinforcing of soft cohesive soils with stone columns. Journal of Ground Engineering, pp.42–49, 1974.

173 The Overseas Coastal Area Development Institute of Japan: English version of technical standards and commentaries for port and harbour facilities in Japan., 2002.

174 Tokimatsu, K. and Yoshimi, Y.: Empirical correlation of soil liquefaction based on SPT N-value and fines content. SOILS AND FOUNDATIONS, Vol.23, No.4, pp.56–74, 1983.

175 Yasuda, S., Harada, K., Niwa, S., Ohkawara, I. and Ogawa, N.: Relationship among SPT N-value, density and K_0 value of silty sand. Proc. of the 36th Annual

Conference of the Japanese Society of Soil Mechanics and Foundation Engineering, pp.1097–1098, 2001 (in Japanese).

176 Hirama, K.: On application of relative density of sand. Proc. of the Symposium on Relative Density and Mechanical Property of Sand, pp.53–56, 1981 (in Japanese).

177 Meyerhof, G.G.: Discussion of session 1. Proc. of the 4th Internal Conference on Soil Mechanics and Foundation Engineering, Vol.3, p. 110, 1957.

Chapter 8

Important Issues on Design Procedures for Clay Ground

8.1 INTRODUCTION

The current design procedures for bearing capacity, stability, earth pressure and horizontal resistance of clay ground are introduced in Chapter 2. In this Chapter, some important issues on the design procedures for clay ground are described to understand the background of the design procedures and suitable choice of design parameters, and to achieve an economical and reasonable design, which includes the choice of shear strength formula, the settlement of floating type improved ground, the strength of sand piles, the stress concentration ratio and the ground deformation due to sand piles installation, etc.

8.2 SHEAR STRENGTH FORMULA

8.2.1 Frequency of case histories

As described in Chapter 2.4, four formulas for evaluating the shear strength of composite ground have been proposed in Japan for slip circle analysis [1–2], which are summarized as follows:

Formula (1):

$$\tau = (1 - as) \cdot (c_0 + kz + \mu_c \cdot \Delta\sigma_z \cdot c_u/p \cdot U)$$
$$+ (\gamma_s \cdot z + \mu_s \cdot \Delta\sigma_z) \cdot as \cdot \tan\phi_s \cdot \cos^2\theta \qquad (8.1a)$$

Formula (2):

$$\tau = (1 - as) \cdot (c_0 + kz)$$
$$+ (\gamma_m \cdot z + \Delta\sigma_z) \cdot \mu_s \cdot as \cdot \tan\phi_s \cdot \cos^2\theta \qquad (8.1b)$$

Formula (3):

$$\tau = (\gamma_m \cdot z + \Delta\sigma_z) \cdot \tan\phi \cdot \cos^2\theta \qquad (8.1c)$$

Formula (4):

$$\tau = (\gamma_m \cdot z + \Delta\sigma_z) \cdot \tan\phi_m \cdot \cos^2\theta \qquad (8.1d)$$

where:
- as : replacement area ratio
- c_0 : shear strength of clay at ground surface (kN/m²)
- c_u/p : shear strength increment ratio
- k : increment ratio of shear strength of clay with depth (kN/m³)
- n : stress concentration ratio

$$n = \frac{\sigma_s}{\sigma_c}$$

- U : average degree of consolidation
- z : depth (m)
- $\Delta\sigma_z$: increment of vertical load intensity (kN/m²)
- γ_s : unit weight of sand pile (kN/m³)
- γ_m : average unit weight of improved ground (kN/m³)

$$\gamma_m = \gamma_s \cdot as + \gamma_c \cdot (1 - as)$$

- θ : inclination of slip circle
- μ_c : stress concentration coefficient of clay ground for external load

$$\mu_c = \frac{\sigma_c}{\sigma} = \frac{1}{1 + (n-1) \cdot as}$$

- μ_s : stress concentration coefficient of sand pile for external load

$$\mu_s = \frac{\sigma_s}{\sigma} = \frac{n}{1 + (n-1) \cdot as}$$

- σ_c : vertical stress on clay ground (kN/m²)
- σ_s : vertical stress on sand pile (kN/m²)
- τ : shear strength of improved ground (kN/m²)
- ϕ : internal friction angle of sand
- ϕ_m : internal friction angle of sand pile

$$\phi_m = \tan^{-1}(\mu_s \cdot as \cdot \tan\phi_s)$$

- ϕ_s : internal friction angle of sand pile

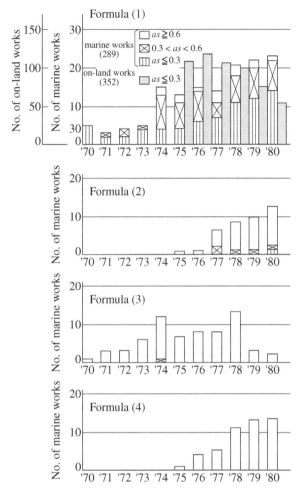

Figure 8.1.   Frequency of case history on selecting design formula for port facilities [3]. (Note the numbering order of Formulas (1) to (4) in this text is different from that in their text.)

Detailed explanations of these formulas were presented by Sogabe [3] and Kanda and Terashi [4]. The selection of design formula is an important issue. Kanda and Terashi [4] studied many case histories on the selection of design formula. Figures 8.1 [3] and 8.2 [4] summarize the frequency of case histories on design formula, which have been applied for design procedures for constructing port and harbor facilities (Note the numbering order of Formulas (1) to (4) in this text is different from that in their text). Figure 8.1 shows that Formulas (1) and (2), where the shear strength component of cohesion and friction is incorporated, have been applied to many case histories irrespective of the replacement area ratio, as. The application of Formula (3) is very limited, and

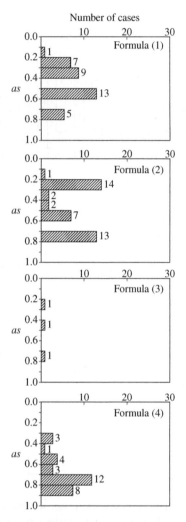

Figure 8.2. Frequency of case histories on design formula [4]. (Note the numbering order of Formulas (1) to (4) in this text is different from that in their text.)

Formula (4), where the shear strength component of cohesion is not incorporated, has been applied to case histories with a relatively high replacement area ratio.

8.2.2 Stress concentration ratio and safety factor

The stress concentration ratio, n, is a significant parameter which influences the shear strength mobilized in sand piles and the increment of shear strength of clay. There are

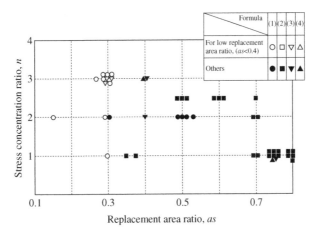

Figure 8.3.   Statistics of relationship between replacement area ratio and stress concentration
ratio [4]. (Note the numbering order of Formulas (1) to (4) in this text is different
from that in their text.)

many accumulated data on the ratio as described in Chapter 7.5. However, these data
are very scattered depending upon many factors such as original and improved ground
conditions, loading condition and type of measurement, so a definitive value of the
ratio has not yet been obtained.

Kanda and Terashi [4] showed statistics of case histories on the stress concentration ratio used in each formula as shown in Figure 8.3. The stress concentration ratio,
n, in the formulas has usually been selected as 1 to 3, while n has been selected as 2 to
3 for improved grounds with a relatively low replacement area ratio.

Bearing capacity of SCP improved ground is evaluated based on assumed
SCP ground conditions by a slip circle analysis with a suitable shear strength formula.
The calculation should be conducted by changing the ground conditions to obtain a
suitable SCP improvement condition (width, depth and replacement area ratio) with
an allowable safety factor. Figure 8.4 shows the statistics of the relationship between
stress concentration ratio and safety factor in Formula (1) [4]. In the figure, the target
and the design safety factors represent an allowable and a calculated safety factor
respectively. The figure indicates that the target safety factor of 1.2 to 1.3 has usually
been adopted in Formula (1) with a stress concentration ratio of 1 to 3. In the figure,
there are three case histories with an extremely high safety factor. These cases are
for improved grounds with the replacement area ratio exceeding 0.7. The allowable
safety factor for slip circle analysis is dependent upon many factors such as type,
size and importance of superstructure, and design stage (during or after construction).
In practical designs for port and harbor facilities, the safety factor of 1.2 or 1.3 is
usually adopted for during construction and at completion of construction,
respectively.

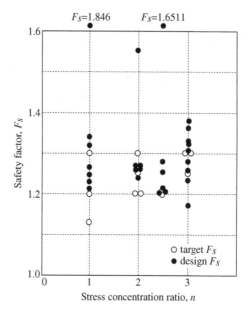

Figure 8.4.   Statistics of case histories on relationship between stress concentration ratio and safety factor in Formula (1) [4]. (Note the numbering order of Formulas (1) to (4) in this text is different from that in their text.)

8.2.3   Effect of design method on safety factor

Figure 8.5 shows the comparison of the safety factors calculated by the stress distribution method and the slice method [4]. These calculations are conducted for a gravel mound type sea revetment and a caisson type sea revetment. The shear strength formulas plotted in the figure are Formulas (1) and (4). The vertical axis of the figure shows the safety factor ratio of the stress distribution method and the slice method. The horizontal axis shows various design parameters. In the figure, the safety factor ratio calculated by changing the stress concentration ratio alone is plotted as open circles or open squares for Formulas (1) and (4) respectively; those by changing the other parameter alone are plotted as different marks.

The figure shows that the safety factor ratio calculated with Formula (4) is higher than that with Formula (1), which indicates that Formula (4) is rather sensitive to the results. This can be explained by the fact that the shear strength component with respect to internal friction influences the results directly in Formula (4). Kanda and Terashi [4] concluded that:

i) the safety factor calculated by the stress distribution method is higher than that by the slice method,

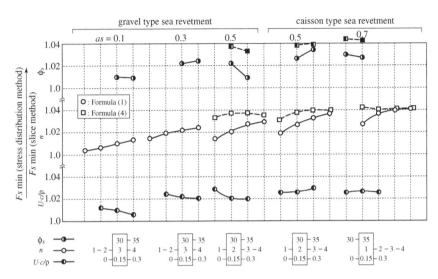

Figure 8.5.   Comparison of safety factors calculated by the stress distribution method and the slice method [4]. (Note the numbering order of Formulas (1) to (4) in this text is different from that in their text.)

ii) the safety factor ratio increases with increasing replacement area ratio, as, and with increasing stress concentration ratio, n.

8.2.4   Effect of design parameters on safety factor

The effect of design parameters (soil parameters) incorporated in the shear strength formulas on the safety factor has been investigated several times [4–6]. The investigations by Kanda and Terashi [4] are presented below.

Figure 8.6 shows the relationship between design parameter and safety factor, $F_{s,min}$, calculated by the slice method. In this calculation, SCP improved grounds with a safety factor of around 1.1 and 1.3 are selected as a basic condition for gravel mound type and caisson type sea revetments. In the figure, the safety factors by changing one of the design parameters alone are plotted by different marks. For example, the safety factors calculated by changing the stress concentration ratio alone are plotted as open circles for the shear strength formula (1), and as open triangles for Formula (2).

In order to compare the effect of design parameters more precisely, the calculations in Figure 8.6 are re-plotted in Figure 8.7 where a safety factor ratio, $F_{s,min}$, is plotted on the vertical axis. The safety factor ratio, $F_{s,min}$, is defined as the ratio of safety factor for each condition to that in the basic condition. The figure shows that the shear strength of clay greatly influences the safety factor ratio for Formula (1), especially in the case of a low replacement area ratio. The effect of the stress concentration

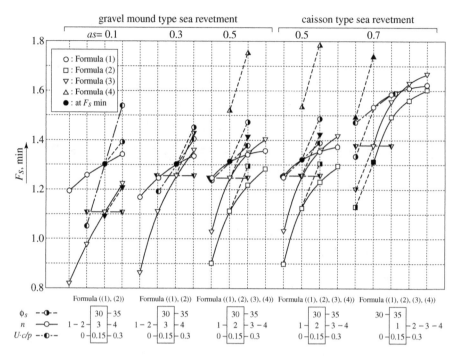

Figure 8.6. Effect of design parameters on safety factor [4]. (Note the numbering order of Formulas (1) to (4) in this text is different from that in their text.)

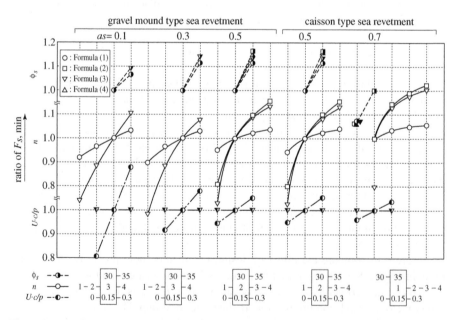

Figure 8.7. Effect of design parameters on safety factor ratio [4]. (Note the numbering order of Formulas (1) to (4) in this text is different from that in their text.)

ratio is almost of the same order in Formulas (2) and (4), but more sensitive for Formula (2) than for Formula (1). This phenomenon can be explained by:

i) the stress concentration ratio influences the shear strengths due to not only the external load but also self weight in Formula (2), but due to the external load alone in Formula (1),

ii) the shear strength increment of clay is also influenced by the stress concentration ratio. With increasing stress concentration ratio, the shear strength mobilized in sand piles increases but the shear strength in clay decreases in Formula (1).

8.3   SETTLEMENT OF FLOATING TYPE IMPROVED GROUND

8.3.1   Amount of settlement

The settlement behavior of fixed type improved ground is described in Chapter 2.5; that of floating type improved ground is described here. As the ground settlement of the floating type improvement is quite complicated, the interaction of the SCP improved layer and soft clay layer underlying it is not yet well understood. The ground settlement can be estimated by summing up the ground settlement taking place in the SCP layer, S_t, and in the soft clay layer, S_u, as shown in Figure 8.8 and Equation (8.2). The settlement in the SCP layer, S_t, is calculated in the same manner as the fixed type improvement with a comparatively small stress concentration ratio. The settlement in the soft layer, S_u, on the other hand, can be estimated by considering the stress distribution as schematically shown in Figure 8.9. The angle of the stress distribution, θ, is not yet clear, but is considered to be $\tan^{-1}(1/2)$ in the design of deep mixing improved ground.

$$S = S_t + S_u$$
$$= \beta \cdot S_0 + S_u \tag{8.2}$$

where:
S : settlement of SCP improved ground (m)
S_t : settlement of SCP improved layer (m)
S_u : settlement of unimproved layer underlying SCP improved layer (m)

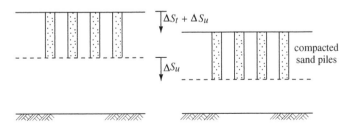

Figure 8.8.   Illustration of settlement of floating type of improved ground.

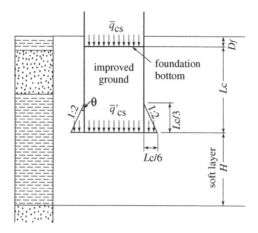

Figure 8.9.   Illustration of stress distribution for floating type Deep Mixing improved ground.

8.3.2   Speed of settlement

Some field settlement measurements of floating type improved grounds were reported by Ogawa and Ichimoto [7]. The consolidation phenomenon of floating type SCP improved ground is similar to that of partially penetrated vertical drains. As there have been few researches on this subject of SCP improved ground, a proposal on the partial penetrating vertical drain method [8] is introduced here, which is applicable to SCP improved ground.

Figure 8.10 illustrates an actual and model of the consolidation phenomenon of partial penetrated vertical drains. In the actual situation, the water in the ground far from the bottom of vertical drains flows vertically upward or downward. The water in the ground close to the bottom of vertical drains, on the other hand, does not flow vertically but flows toward the bottom of drains, which means that the drainage path becomes rather longer than the vertical flow. The degree of consolidation in the unimproved portion is delayed due to the elongation of drainage path. In the proposed calculation, an imaginary horizontal drainage layer is introduced at a depth shallower than the bottom of drains to take into account the above-mentioned phenomenon, and the water in the unimproved portion is assumed to flow vertically to the imaginary drainage layer. Therefore, the thickness of the unimproved portion increases to H' from the original thickness of H. The time settlement relation in the improved ground portion is calculated by the conventional Barron's theory together with the modified coefficient of consolidation, and that in the unimproved portion is calculated by Terzaghi's one-dimensional consolidation theory. The time settlement relation can be calculated by summing up those in the improved and the unimproved portions, which is expressed by Equation (8.3).

$$S_{(t)} = S_{t(t)} + S_{u(t)} \tag{8.3}$$

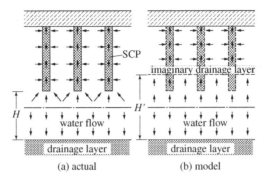

Figure 8.10.   Illustration of actual and model of consolidation phenomenon of partial penetrated vertical drains.

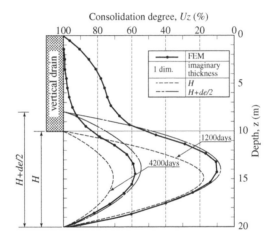

Figure 8.11.   Consolidation degree distribution of partially penetrated vertical drain [8].

where:

$S_{(t)}$: settlement of SCP improved ground at time t (m)

$S_{t(t)}$: settlement taking place in improved portion at time t (m)

$S_{u(t)}$: settlement taking place in unimproved portion at time t (m)

Figure 8.11 shows an example of the consolidation degree distribution of a partially penetrated vertical drain, which is calculated by FEM analyses [8]. In the figure, the calculated values based on the assumption of the imaginary drainage layer are also plotted. Figure 8.11 indicates that the consolidation degree distribution obtained by the proposed method with the thickness of unimproved area of $H+de/2$ coincides well with the FEM analyses, where de is the effective distance of vertical drains.

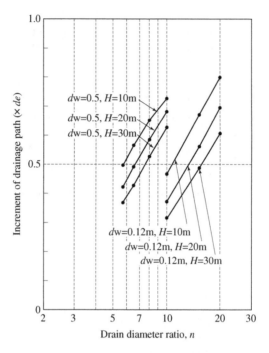

Figure 8.12.   Relationship between the increment of drainage path and drain diameter ratio [8].

The suitable length of drainage path from the bottom of vertical drains to the imaginary drainage layer is influenced by many factors, such as thickness of unimproved portion, diameter of vertical drains and drain interval. Figure 8.12 shows the increment ratio for several conditions, which are obtained as best fitted values by comparing with the FEM analyses [8]. The increment ratio is defined as the drainage path of the imaginary drainage layer to the thickness of unimproved portion. Figure 8.12 shows that the suitable increment ratio increases linearly with increasing drain diameter ratio, $n = de/dw$. It is concluded in the proposal that the increment ratio of 0.5 is applicable with small discrepancy for the case where the thickness of unimproved portion is sufficiently larger than the drain interval, de.

8.4   STRENGTH OF COMPACTED SAND PILE

The strength of compacted sand piles is an important issue especially for application to clay grounds. The strength of compacted sand piles is usually evaluated by the SPT N-value. The continuity of sand piles is also estimated by the distribution of SPT N-value with depth. The frequency of SPT tests for quality assurance is described in Chapter 4.6. The SPT N-value of compacted sand piles is influenced by many factors

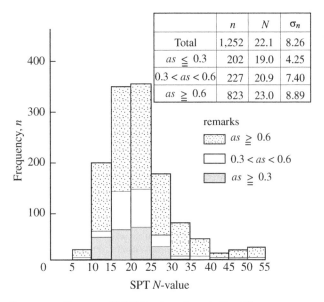

	n	N	σ_n
Total	1,252	22.1	8.26
$as \leq 0.3$	202	19.0	4.25
$0.3 < as < 0.6$	227	20.9	7.40
$as \geq 0.6$	823	23.0	8.89

remarks

$as \geq 0.6$

$0.3 < as < 0.6$

$as \geq 0.3$

Figure 8.13.   Frequency diagram of SPT N-value of sand piles [6].

such as degree and method of compaction, sand pile material, properties of surrounding clay ground, replacement area ratio, etc. The influence of sand material on the SPT N-value is described in Chapter 5 for copper slag material and a mixture of oyster shell and sand.

Figure 8.13 shows accumulated field measurements on the SPT N-value of compacted sand piles [6]. It is found that the SPT N-values of improved grounds with the replacement area ratio of less than 0.3 ranges from 15 to 25 with an average of 19.0. The SPT N-value slightly increases with increasing replacement area ratio, while that for the replacement area ratio exceeding 0.6 shows a relatively high value of 23.0. This indicates that a high replacement area ratio brings high strength to improved ground and high confining pressure, which in turn provides a high SPT N-value.

Figures 8.14a and 8.14b show the relationship between the SPT N-value and cohesion of original clay ground, and replacement area ratio [6]. In Figure 8.14a, the relationship is plotted for several S values, amount of sand per unit depth. The figure shows that the SPT N-value increases with increasing cohesion of clay ground irrespective of S value, and with increasing replacement area ratio, as.

The internal friction angle of sand piles is usually estimated by the SPT N-value in practical designs. There are several proposed relationships between these two parameters as summarized in Table 8.1 and in Figure 8.15 [9]. Among them, the equations proposed by Dunham have been frequently applied for the SCP method as Equations (8.4) to (8.7). However, the effect of overburden pressure is not incorporated in these relationships.

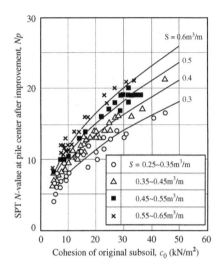

Figure 8.14a.   Effect of cohesion of original ground on SPT N-value of sand piles [6].

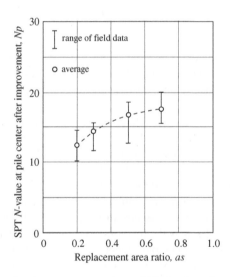

Figure 8.14b.   Effect of replacement area ratio on SPT N-value of sand piles [6].

by Dunham

$$\phi = \sqrt{12 \cdot N} + 25$$
$$\phi = \sqrt{12 \cdot N} + 20 \tag{8.4}$$
$$\phi = \sqrt{12 \cdot N} + 15$$

Table 8.1.   Relationship between SPT N-value and internal friction angle [9].

| SPT N-value | Relative density | | Internal friction angle | |
			by Terzaghi and Peck	by Meyerhof
0 to 4	very loose	0.0 to 0.2	less than 28.5	less than 30
4 to 10	loose	0.2 to 0.4	28.5 to 30	30 to 35
10 to 30	medium	0.4 to 0.6	30 to 36	35 to 40
30 to 50	dense	0.6 to 0.8	36 to 41	40 to 45
higher than 50	very dense	0.8 to 1.0	higher than 41	higher than 45

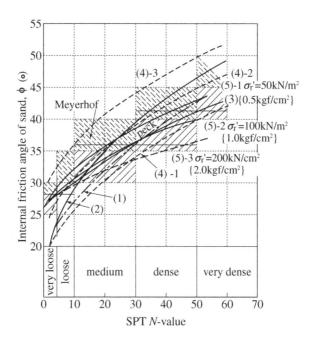

Figure 8.15.   Relationship between SPT N-value and internal friction angle [9].

by Meyerhof
$$\phi = 1/4 \cdot N + 32.5 \qquad (8.5)$$

by Peck
$$\phi = 0.3 \cdot N + 27 \qquad (8.6)$$

by Osaki
$$\phi = \sqrt{20 \cdot N} + 15 \qquad (8.7)$$

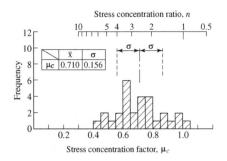

Figure 8.16a.   Stress concentration ratio estimated by shear strength increase measurements [11].

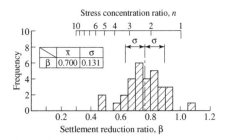

Figure 8.16b.   Stress concentration ratio estimated by ground settlement measurements [11].

8.5   STRESS CONCENTRATION RATIO

Stress concentration ratio is one of significant parameters, which influences bearing capacity and settlement calculation. The ratio is influenced by many factors such as ground condition, strength of sand pile, execution type and type of measurement (see Chapter 7.5). Figures 8.16a and 8.16b show the frequency distribution of stress concentration ratio obtained from field measurements of shear strength increment of clay at Maizuru Port (see Chapter 5.3) [10]. Figures 8.16a and 8.16b show the ratio estimated by measured shear strength increments of clay ground and by back calculating ground settlement measurements respectively. It is found that the estimated stress concentration ratio varies even if it is measured at one site, and differs depending on the type of measurements.

Figures 8.17a and 8.17b show an example of statistical records on stress concentration ratio [6]. Figure 8.17a shows that the stress concentration ratio varies considerably, ranging from 1 to 12 at sites. It is also found that the ratio varies slightly by the type of measurements: the ratio estimated by direct measurements of vertical stresses is the highest value and that estimated by shear strength increments of clay is the lowest value. Figure 8.17b shows the relationship between stress concentration ratio and replacement area ratio, a_s. It is found that the stress concentration ratio varies slightly depending on the replacement area ratio, a_s.

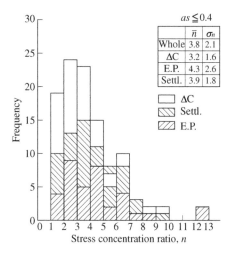

Figure 8.17a.   Frequency distribution of stress concentration ratio obtained from various measurements [6].

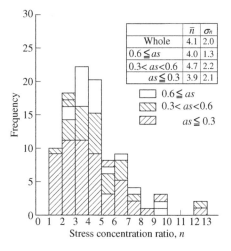

Figure 8.17b.   Frequency distribution of stress concentration ratio obtained from shear strength measurements [6].

8.6   STRENGTH REDUCTION AND RECOVERY OF CLAY

It is well known that clays are subjected to disturbance during installation of vertical drains. Similar to the vertical drain method, clay is subjected to disturbance during sand piles installation. This disturbance effect is dominant in SCP improved grounds

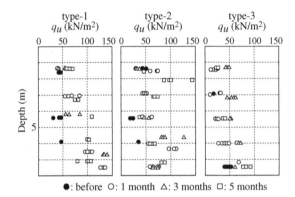

Figure 8.18.   Shear strength distribution with depth [12].

because the replacement area ratio of the SCP method is much higher than that of the vertical drain method. The soil disturbance causes a significant decrease of soil strength, while the soil strength recovers by the dissipation of excess pore water pressure developed during sand piles installation. These phenomena should be precisely evaluated to calculate the bearing capacity and settlement of SCP improved ground more accurately.

Typical field data on the shear strength decrease and its recovery during and after execution is shown in Figure 8.18 [12], where no surcharge load was applied after the execution. Unconfined compressive strength is plotted on the horizontal axis. The figure clearly shows that the soil strength decreases significantly due to sand piles installation but increases very rapidly with elapsed time. The strength increase after one month exceeds unity, which may be caused by the increase of overburden pressure due to ground heaving.

Figures 8.19a and 8.19b show statistics of field measurements on strength reduction and recovery [13]. The figures show the histogram and average strength ratio, defined as the ratio of unconfined compressive strength of improved ground, q_u, to that of the original ground, q_{u0}. Figure 8.19a shows the ratio of clay between sand piles, while Figure 8.19b shows the ratio of the clay periphery of the improved area. Figure 8.19a shows that the unconfined compressive strength just after sand piles installation (within one month) decreases by about 19% on average. However, a larger strength reduction can be seen in marine construction, where the strength decreases to about 60% in the case of the replacement area ratio of less than 0.3. After one month has passed, it is found that the unconfined compressive strength increases more than the original strength, about 11% higher. The strength increases further with elapsed time, where the strength becomes about 120% of the original value at three months after sand piles installation.

In marine construction, the SCP method with a relatively high replacement area ratio has been frequently adopted. This causes soil disturbance of the ground not only

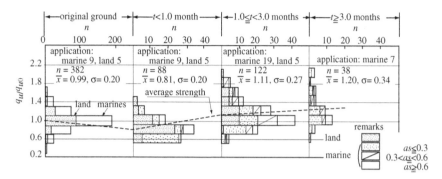

Figure 8.19a.   Strength change of clay in improved area [13].

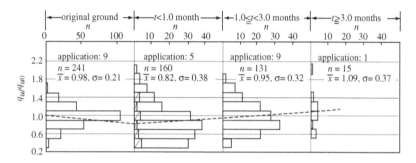

Figure 8.19b.   Strength change of clay periphery of improved area [13].

between sand piles but also in the periphery of the improved area, as shown in Figure 8.19b. The extent of the soil disturbance in the periphery of improved area has been investigated by field measurements. Figure 8.19b shows the statistics of measured data in the periphery of improved area in marine construction. The unconfined compressive strength just after sand piles installation (within one month) decreases by about 18%, which is almost the same order as that between sand piles, as shown in Figure 8.19a. As time elapses, the unconfined compressive strength recovers gradually to 95% of the original strength at one to three months after. The strength then recovers further and becomes higher than the original strength at more than three months after sand piles installation. It should be noted that the process of strength recovery is slower in the periphery rather than between the sand piles. This indicates that the pore water pressure generated during the sand piles installation dissipates rapidly between the sand piles with the help of the drainage function of sand piles. This also emphasizes that the permeability of sand piles and sand mat spreaded on the ground surface should be sufficiently high to assure hydraulic continuity.

   In conclusion, the unconfined compressive strength of the clay between sand piles and periphery of improved area reduces considerably due to sand piles installation, but recovers rapidly and increases higher than the original value about a couple months after installation, provided the sand piles function adequately as a drainage path.

Furthermore, the soil disturbance effect can be neglected in practical design provided it usually takes more than a month after execution to construct a superstructure on SCP improved ground.

8.7   GROUND DEFORMATION DUE TO INSTALLATION OF SAND PILES

8.7.1   Horizontal ground displacement

As some amount of sand is installed into a ground, the ground is generally upheaved and/or deformed horizontally. Figure 8.20 shows statistics of field data on horizontal displacement due to improvement with a relatively low replacement area ratio of 0.1 to 0.3 [14]. These data were measured at on-land construction sites. The magnitude of ground horizontal displacement is dependent upon the geometric condition of improved ground. In applications to sandy ground, horizontal displacements of the order of 20 cm take place at the periphery of the improved ground. In applications to clay and organic grounds, on the other hand, relatively large ground deformations take place, with horizontal displacements of more than 50 cm found in some case records.

8.7.2   Ground heaving

Coefficient of upheaval
As a result of installing some amount of sand into a soft ground, the ground surface can heave to some extent. Figure 8.21 shows a typical case record on the shape of upheaval

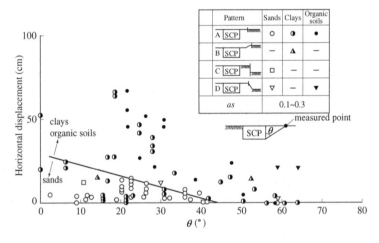

Figure 8.20.   Ground deformation due to SCP execution in on-land construction [14].

ground measured at Maizuru Port where sand piles of 1.6 m in diameter were installed at an interval of 1.7 m (replacement area ratio of 0.7) [15]. The shape and volume of ground heaving is dependent upon many factors such as the replacement area ratio, type of execution machine, construction sequence of sand piles, etc. The upheaved portion was usually excavated to the original ground level to prevent local failure there. However, many researches were carried out to investigate the soil properties of upheaval ground and showed that the shear strength of the upheaval portion reduced greatly just after the execution but rapidly recovered to the original strength or more [15]. This has encouraged the use of the upheaval portion in which sand piles are constructed by a SCP machine to accelerate the strength increase.

Figure 8.22 shows the accumulated field data on the relationship between coefficient of upheaval, μ, and the replacement area ratio, as [3,16–17]. The coefficient of upheaval, μ, is defined as the ratio of the volume of the upheaval portion to the volume of sand introduced. Although there is much scatter in the data, the coefficient

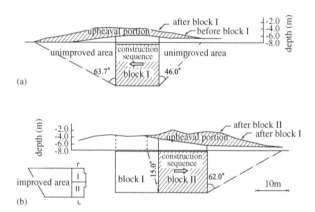

(a)

(b)

Figure 8.21. A typical case record on upheaval ground at Maizuru Port [15].

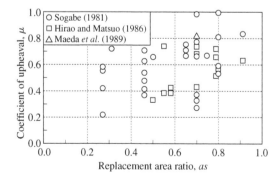

Figure 8.22. Relationship between replacement area ratio and coefficient of upheaval.

of upheaval, μ, increases with increasing replacement area ratio, as. This means that the volume of the upheaval portion increases as a quadratic function with increasing replacement area ratio.

Several equations for predicting the coefficient of upheaval have been proposed as summarized in Equations (8.8a) to (8.8c) [3,18–19] according to accumulated field data. As the unit system is not consistent in the proposed formulation and the constants in these formulas have various units, these formulations are applicable to specific combinations of units for each parameter.

$$\mu = 0.316 \cdot as - 0.028 \cdot L + 0.0037 \cdot q_u + 0.700 \qquad (8.8a)$$

$$\mu = 2.803 \cdot (1/L) + 0.356 \cdot as + 0.112 \qquad (8.8b)$$

$$\mu = 2.477 \cdot (1/L) + 0.400 \cdot as + 0.101 \cdot D + 0.011 \qquad (8.8c)$$

where:
    as : replacement area ratio
    D : diameter of sand pile (m)
    L : sand pile length (m)
    q_u : unconfined compressive strength of original ground at the depth of $L/3$
       (ton/m^2)
    μ : coefficient of upheaval

Figures 8.23a and 8.23b show the relationship between estimated and measured coefficients of upheaval for Equations (8.8a) and (8.8b) respectively. Each equation can predict the measured coefficient with relatively high accuracy for a limited number of field data. However, as described in Chapter 7, Equation (8.8a) has a potential limitation on application to sand piles of length less than 20 m.

Prediction of shape of upheaval portion
As described in the previous section, there are three major methods for predicting the shape of the upheaval portion [16–18]. As each method is provided based on their own field data, there is no universal prediction method applicable to any field condition. Here, the prediction method proposed by Hirao and Matsuo [16], which could be applicable for one direction installation sequence, is exemplified as follows. Figure 8.24 shows a flow chart of predicting the shape of the upheaval portion due to sand piles installation [16]. In the flow, the coefficient of upheaval, μ, is calculated by Equation (8.8b) according to a specific SCP improvement condition.

i) First, the coefficient of upheaval, μ, is evaluated by Equation (8.8b) according to the designed improved ground conditions such as the replacement area ratio, as, width and thickness of improved ground, B and L, and amount of sand to be installed, Vs.

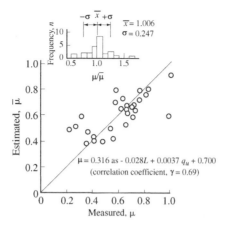

Figure 8.23a.   Relationship between measured and estimated coefficients of upheaval estimated by Equation (8.8a) [14].

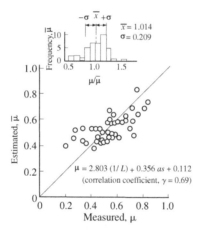

Figure 8.23b.   Relationship between measured and estimated coefficients of upheaval estimated by Equation (8.8b) [14].

ii) The volume of the upheaval portion, V, is calculated by multiplying the amount of sand to be installed and the calculated coefficient of upheaval.

iii) Several shape factors are proposed to characterize the shape of the upheaval portion. Hirao and Matsuo [16] modeled the shape of the upheaval portion as illustrated in Figures 8.25a and 8.25b for one directional and two directional construction sequence respectively, which are characterized by the maximum height, H_{max}, the location at the maximum height, x, the heights at the front and rear edges of improved ground, H_1 and H_2, and the expansion of ground heaving

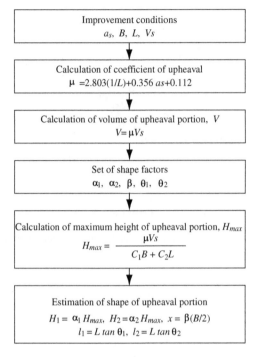

Figure 8.24.   Flow chart of predicting shape of upheaval portion [16].

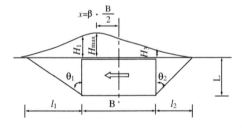

Figure 8.25a.   Modeled shape of upheaval portion for one directional construction sequence [16].

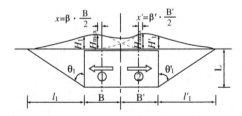

Figure 8.25b.   Modeled shape of upheaval portion for two directional construction sequence [16].

Table 8.2. Shape factors [16].

Shape factor	Description	Value
α_1	The ratio of the height at the front edge of improved ground, H_1, and the maximum height, H_{max}	0.85
α_2	The ratio of the height at the rear edge of improved ground, H_2, and the maximum height, H_{max}	0.4
β	The ratio of the position at maximum height measured from the center of improved ground, x, with respect to half of the improved ground width, $B/2$	0.7
θ_1	Inclination of the upheaval extension at the front side of improved ground	60 degrees
θ_2	Inclination of the upheaval extension at the rear side of improved ground	45 degrees

at front and rear sides, l_1 and l_2. These characteristic parameters are normalized with respect to the shape factors such as the ratio of H_1 and H_{max}, H_2 and H_{max}, inclinations of the upheaval extension, θ_1 and θ_2. The amounts of these values are proposed as shown in Table 8.2 according to the previous case records.

iv) The maximum height of the upheaval portion is calculated by Equation (8.9). This equation is derived from a linear approximation of the upheaval portion.

$$H_{max} = \frac{\mu \cdot Vs}{C_1 \cdot B + C_2 \cdot L}$$
$$C_1 = \frac{1}{2} + \frac{1-\beta}{4}\alpha_1 + \frac{1+\beta}{4} \cdot \frac{1+3\alpha_2}{4} \qquad (8.9)$$
$$C_2 = \frac{1}{2}(\alpha_1 \cdot \tan\theta_1 + \alpha_2 \cdot \tan\theta_2)$$

where:

B : width of SCP improved ground (m)
L : sand pile length (m)
C_1 : constant
C_2 : constant
H_{max} : maximum height of upheaval portion (m)
Vs : volume of sand installed (m³)
α_1 : ratio of H_1 and H_{max}
α_2 : ratio of H_2 and H_{max}
β : ratio of x and $B/2$
μ : coefficient of upheaval
θ_1 : inclination of extension of upheaval portion measured from the bottom of SCP improved ground
θ_2 : inclination of extension of upheaval portion measured from the bottom of SCP improved ground

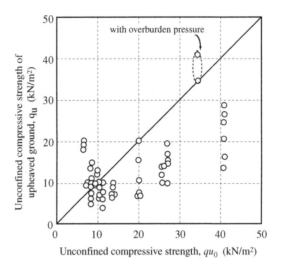

Figure 8.26.   Unconfined compressive strength of upheaval soil and original ground [3].

v) After obtaining the maximum height of the upheaval portion, the shape of the upheaval portion is calculated by Equation (8.10).

$$H_1 = \alpha_1 \cdot H_{max} \tag{8.10a}$$

$$H_2 = \alpha_2 \cdot H_{max} \tag{8.10b}$$

$$x = \beta \cdot \frac{B}{2} \tag{8.10c}$$

$$l_1 = L \cdot \tan \theta_1 \tag{8.10d}$$

$$l_2 = L \cdot \tan \theta_2 \tag{8.10e}$$

Shear strength of upheaved soil
According to the accumulated field data, the properties of upheaval soil can be assumed to be the same as that at a shallow depth of ground. Figure 8.26 shows the relationship between the unconfined compressive strength, q_u, of upheaval soil and that of the original ground at a shallow depth [3]. The figure shows that the q_u value is almost the same as that of the original soil except at relatively high strength exceeding about 20 kN/m².

REFERENCES

1 Ministry of Transport: Technical standards and commentaries for port and harbour facilities in Japan. Ministry of Transport, Japan, 1999 (in Japanese).

2   The Overseas Coastal Area Development Institute of Japan: English version of technical standards and commentaries for port and harbour facilities in Japan. 2002.

3   Sogabe, T.: Technical subjects on design and execution of sand compaction pile method. Proc. of the 36th Annual Conference of the Japan Society of Civil Engineers, III, pp.39–50, 1981 (in Japanese).

4   Kanda, K. and Terashi, M.: Practical formula for the composite ground improved by sand compaction pile method. Technical Note of the Port and Harbour Research Institute, No.669, pp.52, 1990 (in Japanese).

5   Kuno, G. and Nakayama, J.: The improvement effect of Compozer Method to stability of embankment. Journal of the Japanese Society of Soil Mechanics and Foundation Engineering, 'Tsuchi-to-Kiso', Vol.16, No.12, pp.11–19, 1968 (in Japanese).

6   Ichimoto, E. and Suematsu, N.: Actual practices and subjects of sand compaction pile method (3) – summary – Journal of the Japanese Society of Soil Mechanics and Foundation Engineering, 'Tsuchi-to-Kiso', Vol.31, No.5, pp.83–90, 1983 (in Japanese).

7   Ogawa, M. and Ichimoto, E.: Application of vibro-Compozer method to cohesive ground. Journal of the Japanese Society of Soil Mechanics and Foundation Engineering, 'Tsuchi-to-Kiso', Vol.11, No.3, pp.3–9, 1963 (in Japanese).

8   Shiomi, M.: Consolidation settlement control of improved ground by vertical drain method. Ph.D. Theses, 1998 (in Japanese).

9   Fujita, K.: Standard penetration test. Comments and Applications of Soil Survey Test Results, The Japanese Society of Soil Mechanics and Foundation Engineering, pp.29–75, 1968 (in Japanese).

10 Okada, Y., Yagyuu, T. and Sawada, Y.: Field loading test of SCP improved ground with low replacement area ratio. Journal of the Japanese Society of Soil Mechanics and Foundation Engineering, 'Tsuchi-to-Kiso', Vol.37, No.8, pp.57–62, 1989 (in Japanese).

11 Yagyu, T. and Yukita, Y.: Field rupture test of soil improved by sand compaction piles with low sand-replacement ratio in sea. Proc. of the 24th Annual Conference of the Japanese Society of Soil Mechanics and Foundation Engineering, pp.1891–1894, 1989 (in Japanese).

12 Matsuo, M., Kimura, M., Nishio, Y. and Ando, H.: Development of soil improvement method using construction waste soil. Journal of Construction Management and Engineering, Japan Society of Civil Engineers, No.567/3–35, pp.237–248, 1997 (in Japanese).

13 Ichimoto, E.: Practical design method and calculation of sand compaction pile method. Proc. of the 36th Annual Conference of the Japan Society of Civil Engineers, III, pp.51–55, 1981 (in Japanese).

14 Japanese Society of Soil Mechanics and Foundation Engineering: Soil improvement methods – survey, design and execution-. The Japanese Society of Soil Mechanics and Foundation Engineering, 1988 (in Japanese).

15   Hirao, H. and Matsuo, M.: Study on upheaval ground generated by sand compaction piles. Journal of Geotechnical Engineering, Japan Society of Civil Engineers, No.364/3–4, pp.169–178, 1985 (in Japanese).

16   Hirao, H. and Matsuo, M.: Study on characteristics of upheaval part of cohesive ground caused by soil improvement. Journal of Geotechnical Engineering, Japan Society of Civil Engineers, No.376/3–6, pp.277–285, 1986 (in Japanese).

17   Maeda, S., Takai, T. and Fukute, T.: Shape and properties of the upheaval of cohesive soil improved by compacted sand piling method. Journal of Construction Management and Engineering, Japan Society of Civil Engineers, No.403/6–10, pp.55–63, 1989 (in Japanese).

18   Shiomi, M. and Kawamoto, K.: Prediction of ground heave associated with the installation of sand compaction piles. Proc. of the 21st Annual Conference of the Japanese Society of Soil Mechanics and Foundation Engineering, pp.1861–1862, 1986 (in Japanese).

19   Hirao, H.: Characteristics and beneficial usage of upheaved ground due to sand piles installation. Ph.D. Theses, pp.150, 1996 (in Japanese).

Chapter 9

Important Issues on Design Procedure for Sandy Ground

9.1 INTRODUCTION

The current design procedures for liquefaction of sandy ground are introduced in Chapter 3. In this Chapter, some important issues on the design procedure for sandy ground are described to understand the background of the design procedures and suitable choice of design parameters, and to achieve an economical and reasonable design which includes the evaluation of design procedures, relationship between relative density and SPT N-value, effect of horizontal stress, etc.

9.2 EVALUATION OF DESIGN PROCEDURE

Figures 9.1a and 9.1b show the evaluation of liquefaction resistance of SCP improved ground, which is based on accumulated field data as summarized in Table 9.1 [1]. Figure 9.1 shows the SPT N-value distributions of sandy grounds with depth, whose fines contents were less than 20%. Figure 9.1a shows those of the original ground and Figure 9.1b shows those of the improved ground where no liquefaction took place during an earthquake. The SPT N-values plotted in the figure were measured within one to four weeks after the improvement, and were normalized for an effective overburden pressure of $100 \, \text{kN/m}^2$ ($1 \, \text{kgf/cm}^2$) by Equation (9.1) [2].

$$N_1 = \frac{170 \cdot N}{\sigma'_v + 70} \tag{9.1}$$

where:
  N : measured SPT N-value
  N_1 : normalized SPT N-value
  σ'_v : overburden pressure (kN/m^2)

By comparing these two figures, it is found that the normalized SPT N-values, N_1, after improvement are 10 to 20 higher than the original values. In the figures, the normalized critical SPT N-values are also plotted as a broken line. The critical SPT

N-values are obtained from the relationship between SPT N_1-value and liquefaction potential [3], and from the cyclic shear strength ratio calculated by Equation (9.2). The figures clearly show that almost all SPT N_1-values of the original ground are lower than the critical values, but many of the SPT N_1-values of improved grounds are

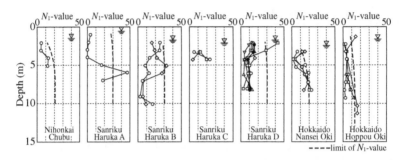

Figure 9.1a.   Normalized SPT N-value distribution with depth before SCP improvement [1].

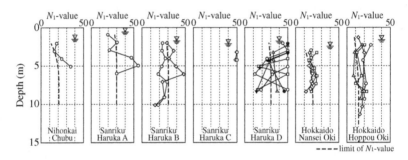

Figure 9.1b.   Normalized SPT N-value distribution with depth after SCP improvement [1].

Table 9.1.   Property of earthquake and SCP improvement pattern for accumulated field data in Figure 9.1 [1].

Name		Max. accel. (gal)	Number of borings	Improvement pattern
Nihonkai Chubu		200	1	2.0 m equilateral triangular
Sanriku Haruka	A	440	1	1.5 m square
	B	600	2	1.9 m equilateral triangular
	C	600	3	1.9 m equilateral triangular
	D	600	6	1.9 m equilateral triangular
Hokkaido Nansei Oki		220	3	1.5 m square
Hokkaido Hoppou Oki		200	4	1.7 m square

higher than the critical values. This indicates that the critical SPT N_1-value distribution curve with depth coincides well with the field measurements.

$$\frac{\tau_{max}}{\sigma'_v} = (1 - 0.015 \cdot z) \cdot \frac{\alpha_{max}}{g} \cdot \frac{\sigma_v}{\sigma'_v} \tag{9.2}$$

where:
   g : acceleration due to gravity (gal)
   z : depth (m)
   α_{max} : maximum acceleration at ground surface (gal)
   σ_v : overburden pressure (kN/m^2)
   σ'_v : effective overburden pressure (kN/m^2)
   τ_{max} : maximum shear stress (kN/m^2)

There are some data in Figure 9.1b which are lower than the critical SPT N_1-value. In order to clarify the phenomenon, the field data in Figure 9.1a are re-plotted in Figure 9.2 as the relationship between normalized SPT N-value and dynamic shear stress ratio, τ_{max}/σ'_v [1]. In the figure, the critical SPT N_1-value proposed by Matsuo [3] is plotted together. For the original ground as shown in Figure 9.2a, many field data are plotted on the left-hand side of the critical value (the normalized SPT N-value of the original ground is less than the critical value), which indicates that the original ground is expected to liquefy during an earthquake. This was confirmed by field observation where some liquefaction phenomena were observed in the surroundings of the SCP improved area. For the improved ground as shown in Figure 9.2b, many

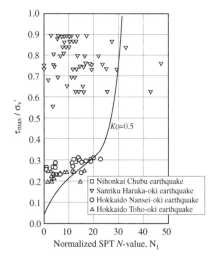

Figure 9.2a.   Relationship between normalized SPT N-value and shear stress ratio before SCP improvement [1].

field data are plotted on the right-hand side of the critical value. The full circle and triangle marks in the figure indicate that field measurements lower than the critical value are encountered at depths of more than 3 m continuously. These measurements are close or lower than the critical value.

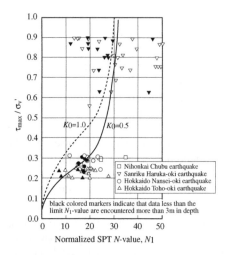

Figure 9.2b.   Relationship between normalized SPT N-value and shear stress ratio after SCP improvement [1].

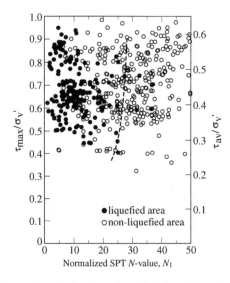

Figure 9.3.   SPT N_1-value at liquefied and non-liquefied sites during the 1995 Hyogoken-nambu earthquake [4].

These figures indicate that SCP improved ground has higher liquefaction resistance than that expected by the increase of SPT N-value alone. This can be explained by an increase of ground stiffness, an increase of horizontal stress and/or a high permeability of sand piles [1].

Yasuda *et al.* [4] also presented a similar figure on SPT N-values at liquefied and non-liquefied sites during the 1995 Hyogoken-nambu earthquake in Figure 9.3. The full circles in the figure represent the SPT N_1-values measured at liquefied sites, while the open circles represent those measured at non-liquefied sites. The figure indicates that grounds with SPT N_1-value exceeding about 25 do not liquefy even in an earthquake as large as the Hyogoken-nambu earthquake. The broken line in the figure indicates the critical SPT N_1-value calculated by Equation (9.1). Although there is much scatter in the measured data, it is found that the critical SPT N_1-value is reasonable for evaluating the liquefaction potential.

9.3   EVALUATION OF IMPROVEMENT EFFECT BASED ON COMPOSITE GROUND

The SPT N-values measured at the center of sand piles are shown in Figure 9.4 for application to sandy ground [1]. The figure shows that the SPT N-values are relatively high at around 23 to 29 on average irrespective of the fines content of the original ground.

The SPT N-values between sand piles, on the other hand, are relatively small and decrease with distance from sand piles, as shown in Figures 9.5a and 9.5b [5]. These data were measured by Swedish cone penetration tests and cone penetration tests, respectively. It is found that the measured ground strength between sand piles decreases very rapidly with distance from sand piles in the case of ground with fines content, Fc, of less than 20%. However, the strength of ground with fines content ranging from

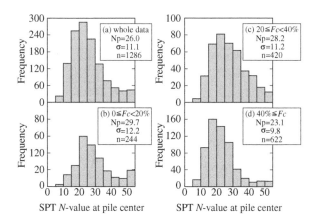

Figure 9.4.   SPT N-values at sand pile center [1].

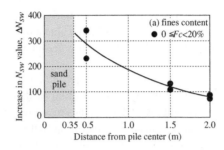

Figure 9.5a.   Strength measurements between sand piles by Swedish cone penetration tests [5].

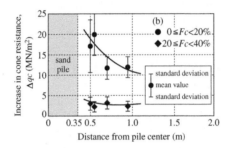

Figure 9.5b.   Strength measurements between sand piles by cone penetration tests [5].

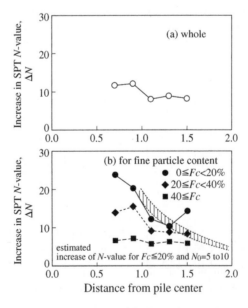

Figure 9.6.   Relationship between increment of SPT N-value and distance from pile center [1].

20% to 40% is quite small value and almost uniform between sand piles. A similar phenomenon can be seen in other field measurements as shown in Figure 9.6 [1].

The improvement effect is usually evaluated by the SPT N-value at the ground between sand piles, where the least improvement effect is achieved. These data indicate that the current design procedure, in which the improvement effect is evaluated at the furthest point from the sand pile, gives a very conservative result for the improvement effect.

As described in Figure 9.1, there are some case histories where SCP improved grounds do not liquefy during an earthquake even if their liquefaction potential is less than the critical one. This phenomenon can be explained by the conservativeness of the evaluation, the effect of horizontal stress and the composite ground effect.

9.4   RELATIONSHIP BETWEEN RELATIVE DENSITY AND SPT N-VALUE

As described in the previous section, the relative density of ground is one of the key parameters in the design. The relative density can be measured directly on an undisturbed specimen of ground. However, undisturbed sampling of sandy ground was and is still difficult and quite expensive. In practical design and construction, the relative density of ground is usually estimated by the SPT N-value. The relationship between the relative density and the SPT N-value was proposed by several researchers such as Terzaghi and Peck, Meyerhof [6], Dunham, and Osaki $et\ al$. After recognizing that the

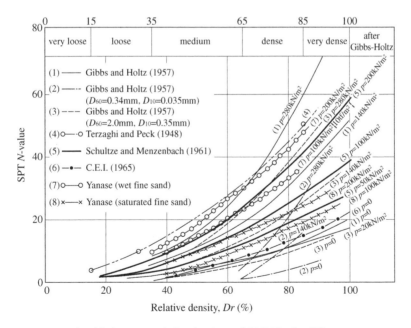

Figure 9.7.   Relationship between relative density and SPT N-value [9].

SPT N-value was influenced by an effective overburden pressure, Gibbs and Holtz [7] and Schultze and Menzenbach [8] proposed another relationship to incorporate the effect.

Figure 9.7 summarizes the relationships between the relative density and the SPT N-value proposed by several research engineers [9]. The figure shows that the SPT N-value increases with increasing relative density of ground and an effective overburden pressure. However, there is much scatter for each soil and each overburden pressure. Among them, the relationship measured by Gibbs and Holtz [7] is adopted for the design procedure B proposed by Ogawa and Ishidou [10] as described in Chapter 3.2.

9.5   EFFECT OF HORIZONTAL STRESS INCREASE

As explained in Chapter 6, sand installed into the ground is subjected to vibration or static force and expands in diameter. The horizontal stress in a ground, in turn, becomes high during sand piles installation. Figure 9.8 shows a typical example of field data on static earth pressure coefficient, K_0, that were measured by bore hole horizontal loading tests [5]. In the figure, the sandy grounds were compacted by one of two techniques: the vibrating compaction technique or the static compaction technique (non-vibrating compaction technique). The figure clearly shows that the static earth pressure coefficient, K_0, increases from the original value of around 0.5 to higher than 1.0. Some data marked by double circles in the figure were measured two years after sand piles installation, which shows negligible reduction of K_0 value even after two years.

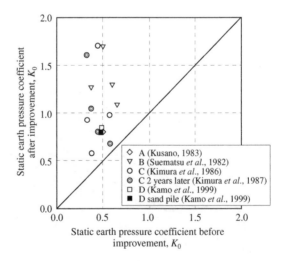

Figure 9.8.   Static earth pressure coefficients before and after SCP improvement [5].

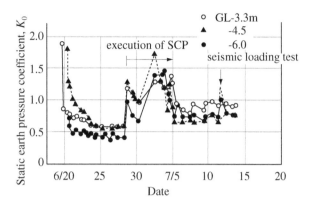

Figure 9.9.   Static pressure coefficient changes during and after SCP improvement [5].

Figure 9.9 shows the change in the static earth pressure coefficient, K_0, after improvement [5]. It can be seen that the coefficient, K_0, increases higher than 1.0 during the SCP improvement and decreases after the improvement. However, the coefficient after the improvement is around 0.8, which is still higher than the original value of around 0.5. Similar field data was obtained silty grounds and sandy grounds [11].

It is well known that the liquefaction resistance of ground is influenced by the static coefficient of horizontal stress [11]. The fact that field measured data show an increase in horizontal stress due to sand piles installation indicates that quality assurance based on the SPT N-value alone can underestimate the liquefaction resistance of improved ground.

9.6   PORE WATER PRESSURE CHANGE DURING INSTALLATION OF SAND PILES

Figure 9.10 shows the pore water pressure change in a sandy ground during sand piles installation. The sandy ground was improved by sand piles with a diameter of 70 cm and square pattern of 1.3 m intervals, which corresponded to the replacement area ratio, as, of 0.228. The pore water pressure was measured at the sandy ground between sand piles, as shown in the upper part of Figure 9.10, and was normalized with respect to the overburden pressure [5].

The figure shows that the excess pore water pressure ratio, $\Delta u/\sigma_v'$, increases to about unity during penetration of a casing pile. During sand piles installation by the vertical vibrating compaction technique, the excess pore water pressure ratio increases to around 0.5. Thereafter, the pore water pressure dissipates very rapidly. This phenomenon indicates that the sandy ground is almost liquefied due to the vibration during penetration of the casing pipe and installation of sand piles, which causes an increase in density of the ground.

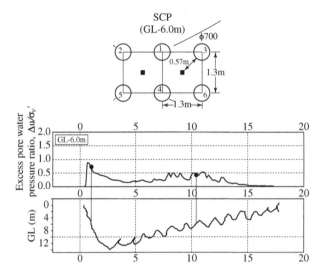

Figure 9.10.   Pore water change during sand piles installation [5].

REFERENCES

1   Ohbayashi, J., Harada, K., Yamamoto, M. and Sasaki, Y.: Evaluation of liquefaction resistance of compacted ground. Proc. of the Symposium on Earthquake Engineering, pp.1411–1416, 1998 (in Japanese).

2   Tokimatsu, K. and Yoshimi, Y.: Empirical correlation of soil liquefaction based on SPT N-value and fines content. SOILS AND FOUNDATIONS, Vol.23, No.4, pp.56–74, 1983.

3   Matsuo, O.: Liquefaction Resistance of sandy soils. Proc. of the 31st Annual Conference of the Japanese Society of Soil Mechanics and Foundation Engineering, pp.1035–1036, 1996 (in Japanese).

4   Yasuda, S., Tsubota, K., Nishikawa, O., Asaka, H. and Naitoh, F.: SPT N-values at liquefied and non-liquefied sites during the 1995 Hyogoken-nambu earthquake. Proc. of the 31st Annual Conference of the Japanese Society of Soil Mechanics and Foundation Engineering, pp.1225–1226, 1996 (in Japanese).

5   Ohbayashi, J.: Study on damage mechanism and countermeasures of embankment due to liquefaction. Ph.D. Theses, 2001 (in Japanese).

6   Meyerhof, G.G.: Penetration tests and bearing capacity. Proc. of the American Society of Civil Engineers, Journal of Soil Mechanics and Foundations Division, pp.1–19, 1956.

7   Gibbs, H.J. and Holtz, W.G.: Research on determining the density of sands by spoon penetration testing. Proc. of the 4th Internal Conference on Soil Mechanics and Foundation Engineering, Vol.1, pp.35–39, 1957.

  8 Schultze, E. and Menzenbach, E.: Standard penetration test and compressibility of soils. Proc. of the 5th Internal Conference on Soil Mechanics and Foundation Engineering, Vol.1, pp.527–532, 1961.

  9 Fujita, K.: Standard penetration test. Comments and Applications of Soil Survey Test Results, The Japanese Society of Soil Mechanics and Foundation Engineering, pp.29–75, 1968 (in Japanese).

 10 Ogawa, M. and Ishidou, M.: Application of 'compozer method' for sandy ground. Journal of the Japanese Society of Soil Mechanics and Foundation Engineering, 'Tsuchi-to-Kiso', Vol.13, No.2, pp.77–81, 1965 (in Japanese).

 11 Kusano, I.: Actual practices and subjects of sand compaction pile method (3) – improvement effect of river dike and dynamic response. Journal of the Japanese Society of Soil Mechanics and Foundation Engineering, 'Tsuchi-to-Kiso', Vol.31, No.4, pp.79–85, 1983 (in Japanese).

 12 Ishihara, K., Iwamoto, S., Yasuda, S. and Takatsu, H.: Liquefaction of anisotropically consolidated sand. Proc. of the 9th International Conference on Soil Mechanics and Foundation Engineering, pp.11–15, 1977.

Index